T0073298

Dynamic Management of Sustainable Development

Zigurds Krishans · Anna Mutule ·
Yuri Merkuryev · Irina Oleinikova

Dynamic Management of Sustainable Development

Methods for Large Technical Systems

 Springer

Zigurds Krishans
Laboratory of Power Systems
  Mathematical Modelling
Institute of Physical Energetics
Aizkraukles 21
LV-1006 Riga
Latvia
e-mail: olegsk@edi.lv

Anna Mutule
Laboratory of Power Systems
  Mathematical Modelling
Institute of Physical Energetics
Aizkraukles 21
LV-1006 Riga
Latvia
e-mail: amutule@edi.lv

Yuri Merkuryev
Department of Modelling and Simulation
Riga Technical University
Kalku Street 1
LV-1658 Riga
Latvia
e-mail: merkur@itl.rtu.lv

Irina Oleinikova
Laboratory of Power Systems
  Mathematical Modelling
Institute of Physical Energetics
Aizkraukles 21
LV-1006 Riga
Latvia
e-mail: irina@edi.lv

ISBN 978-0-85729-055-7   e-ISBN 978-0-85729-062-5

DOI 10.1007/978-0-85729-062-5

Springer London Dordrecht Heidelberg New York

British Library Cataloguing in Publication Data
A catalogue record for this book is available from the British Library

Library of Congress Control Number: 2010936450

Cover design: eStudio Calamar, S.L./Figueres

Printed on acid-free paper

Springer is part of Springer Science+Business Media (www.springer.com)

Preface

Optimisation of large technical systems development makes it necessary to apply systems analysis and dynamic multi-step optimisation methods to observe reciprocal interconnections of system elements over time and space. A characteristic feature of systems analysis is that for selecting an optimal solution, sophisticated systems must be investigated within the development process. To realize this kind of approach, modelling methods of the existing system development have to be used. They must adequately reflect real systems characteristics and perform system technical, economic and ecological criteria calculation. Besides, effective optimization methods are required to solve optimization tasks with discrete variables.

The authors work at the Laboratory for Power Systems Mathematical Modelling (PSMM) of Institute of Physical Energetics, established in 1969. The sphere of scientific research of the laboratory is dynamic methods for electric power systems development optimization.

Initially, research work embraced medium voltage distribution networks; then it was extended to transmission grid and generation systems. In the nineties of the 20th century low voltage network dynamic development optimizations models were also created and development optimization models of generation systems and medium voltage distribution networks were dispersed. The research was conducted in line with practical problems solution in Latvia, former Soviet Union, Baltic States and European Union, in cooperation with power systems managers and designers.

In this book, with system sustainability we identify the fact that up to now in designing process of electrical power systems not enough attention has been drawn to consequences of the made decisions that must not be ignored. Taking into consideration the above-mentioned, we suggest using dynamic methods in system development sustainability analysis, which allows observing development process and far perspective.

Historically, research focused on dynamic methods elaboration was already initiated in the middle of the 20th century, when R. Bellman worked out his dynamic programming method. Still dynamic programming in direct way may only be used for large systems development tasks if the number of alternative development actions is not considerable. Then, increasing variable number phe-nomenon appeared which R.Bellman named as dynamic programming "curse"—the expo-nential growth of calculation time. The specialists of Laboratory for Power Systems Mathematical Modelling (PSMM) have been working to overcome and clear this "curse" and have developed a new original method—"optimal initial states method" (OIS), which provides opportunity to consider up to 50 alternative development actions and solve real large technical systems develop-ment optimization tasks. OIS allows one to comprehensively motivate decisions, through modelling situation and its created consequences, thus evidently helping to select an optimal option. In case of electric power systems, nowadays the decisions must often be made at the state governmental level and under infor-mation uncertainty conditions employing fast operable information technologies.

The purpose of the book is to present Laboratory's valuable long-term expe-rience through elaboration of electric power systems development optimization methods in order to compile the basic principles of these methods and experience realization in practice, so that it might also be used for management of other large technical systems sustainable development with dynamic optimization methods.

The book is intended for engineering scientific and IT researchers, various systems management and operation control specialists and engineering technical staff, systems programmers, as well as for faculty lecturers and students of tech-nical higher educational establishments.

The book is composed of ten Chapters which are divided into three parts:

Basic concepts and definitions—Chapters 1, 2 and 3 (authors Z. Krishans and I. Oleinikova);

Family of optimal initial state methods—Chapters 4, 5, 6 and 7 (authors Z. Krishans and A. Mutule);

Management technology for LTS sustainable development—Chapters 8, 9 and 10 (authors Z. Krishans and Y. Merkuryev).

The first three Chapters review general mathematical basics of a dynamic model of large technical systems sustainable management. Chapters 4 and 5 consider the basic principles of dynamic development optimization methods advancement (optimization area extension). Chapters 6 and 7 examine a set of opti-mal initial states methods (OIS), considered by the authors as the best system for development optimization. Dynamic systems development optimization methods are discussed in detail in Chapters 2 to 7. Chapters 8 to 10 provide a review of in-formation technology for electric power systems, transmission net-work and generation sources sustainable development management.

The aforementioned methods have been elaborated in the Institute of Physical Energetics, Laboratory for Power Systems Mathematical Modelling, for electric power systems development planning, including also problem solution on sus-tainable development management. We consider that these methods may also be em-ployed for other systems sustainable development management. Therefore, we suggest creating dynamic development management systems in the systems with similar development problems. This shall be done now to ensure that the decisions made nowadays be adequate and viable for next generations.

We would be grateful to book readers for all remarks and recommendations. Our contact address is: Laboratory for Power Systems Mathematical Modelling, Institute of Physical Energetics, 21 Aizkraukles Street, LV-1006, Riga, Latvia.

Riga, Latvia, March 2010

Zigurds Krishans
Anna Mutule
Yuri Merkuryev
Irina Oleinikova

Acknowledgements

Authors expressed gratitude to Senior Editorial Assistant (Engineer-ing) of publishing house *Springer* Claire Protherough who gave great assistance in working on this monograph issue.

The material summarised in the monograph represents a long-term research of the Laboratory for Power Systems Mathematical Modelling (PSMM) in the field of dynamic systems. The monograph came into being through the labours of a number of people. The research is conducted in close co-operation with energy power supply enterprises and numerous power supply systems designing institutions where software programs of the LPMM are employed. We also co-operate with technical universities and research institutions that study energy power system development planning related problems.

We appreciate the assistance of our colleagues who have considerably contributed to our research.

First of all we thank both former and current researchers of the Laboratory for PSMM: *Dr.sc.ing.* Paegle Omars, *Dr.hon.c.* Dale Voldemars, *Dr.sc.ing.* Abramova Halina, Greivule Inara, *Dr.sc.ing.* Oleinikova Lidija, Kalpina Anna, Bumane-Luse Inguna and *B.Sc.* Kochukov Oleg.

We would also like to give a special acknowledgement to our colleagues from

- Latvian energy power supply enterprise (Latvenergo): Kruskops A., Lapinskis V. and *Dr.sc.ing.*, prof. Barkans J.;
- power supply system designing institutions: *Dr.habil.sc.ing.* Seiliger A.N., *Dr.sc.ing.* Malkin P.A., Kusnetsova O.N., *Dr.habil.sc.ing.* Jershevich V.V., *Dr.sc.ing.* Abramenkova N.A., Erhards E., Inde J. and Jansons V.;
- technical universities: *prof.* Bubenko J., *prof.* Andersons G., *prof.* Venikov V.A., *prof.* Stroev V.A., *prof.* Glazunov A.A., *prof.* Arzamascev D.A., *Dr. habil.sc.ing.* Mizin A.L., *Dr.sc.ing.* Buslova N.V. and *Dr.sc.ing.* Blok V.M.;
- research institutions: *Dr.habil.sc.ing.* Voropai N.I., *Dr.habil.sc.ing.* Belajev L.S., *Dr.habil.sc.ing.* Arion V.D. and *Dr.habil.sc.ing.* Zhuravlev V.G.

Contents

Abbreviations

CHP	cogeneration heat plant
D-action	development action
D-plan	development plan
D-process	development process
D-state	development state
D-step	development step
HPP	hydropower plant
IRR	Internal Profit Rate
IT	information technology/technologies
LTS	large technical system
LTSs	large technical systems
m.u.	monetary units
NPP	nuclear power plant
NPV	Net Present Value
OIS	optimal initial states
PB	Pay Back Period
PSMM	Power Systems Mathematical Modelling
r.u.	related units
sub./st.	substation
UDM	unit diagonal matrix
WPP	wind power plant

List of Terms and Methods

List of Figures

List of Tables

Chapter 1
Problem Statement

Abstract This chapter describes the basics of large technical system (LTS) development management: the problems of optimal development plan (D-plan) selection, LTS sustainable development methods, and LTS development optimization criteria. This chapter explains that LTS development process (D-process) is dynamic—the creation or reconstruction of system objects is performed in different locations of the system and at different time moments. The relative efficiency of D-plans in D-process is variable—it means that decision has to be taken considering long-time period. In real tasks, D-plan total can reach up to 10^{20}, due to those new methods of LTS optimization have to be developed. LTS sustainable development management is sophisticated activity because of uncertainty of future. To overcome this condition, in this chapter, we come forward with dynamic management methods. This chapter describes LTS development optimization multi-criteria problem, as well as presents new solutions of this problem.

1.1 Key Points of Large Technical Systems Sustainable Development Management

The general structure of large technical system (LTS) is shown in Fig. 1.1. Usually, the structural scheme can be represented as a graph with nodes and transport network. The nodes are subdivided into production and consumption nodes. The nodes are connected with transport network links. Let us use the concept definition *system elements* to denote nodes and network links with.

LTSs are steadily developing. Generation nodes capacity and location is being altered. In addition, the figure of consumption nodes volume is a changeable figure; it may either increase or decrease. Similarly, parameters of transport network links are changing periodically—new links are constructed and existing least-loaded links are removed.

Z. Krishans et al., *Dynamic Management of Sustainable Development*, DOI: 10.1007/978-0-85729-062-5_1, © Springer-Verlag London Limited 2011

Fig. 1.1 Structural scheme
of LTSs

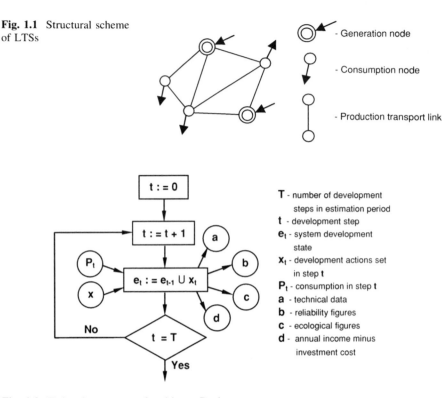

Fig. 1.2 Estimation process of multi-step D-plan

The changes in consumption cause modifications of system structure and parameters that result in new consumption nodes establishment and system elements obsolescence. That defines the necessity to reconstruct or to eliminate certain elements of the existing system.

The D-process of LTS is managed by system management executive representatives and respective management departments as well as by engineering and technical personnel and design organizations. A large number of decisions must be made in this management process concerning the capital investments required and these decisions must be thoroughly substantiated.

In real conditions, the tasks to be solved are rather diversified. In many cases, the task may be solved using just the experience gained, so the solution is visible.

But still there are a lot of cases in the sustainable D-process of LTS when prospective D-plans are multiple and diversified; therefore, to find and determine the optimal D-plan is a very sophisticated task. Only by applying the specialized methods, it is possible to reach the utmost results.

System development optimization process is shown schematically in Fig. 1.2.

The development problems of the existing functioning systems are becoming ever more complicated, which enlarges the number of prospective D-plans and calculation scope for estimation of each D-plan [1, 2].

Rather often, in making an optimal decision, it is indispensable to review and to estimate the system D-process for the longer time period not for particular development "horizon", as the decisions made in primary state may essentially affect system further development. In case if transport network consists of two levels—distribution and transmission, the development of both network complexes must be estimated and reviewed simultaneously.

For D-plan optimization, it is indispensable to apply dynamic optimization methods which are considered in detail in Chaps. 4–6 of the book.

1.2 Criteria of LTS Sustainable Development Management

Optimization incorporates the methods and algorithms, by which the maximum for objective function F is determined, usually with many variables, observing supplemental conditions, including limitations which are expressed by equations or inequations. In tasks on technical systems, development variables are development actions (D-actions), but technical criteria are limitations, such as links maximal capacity, etc. Usually, for development optimization tasks multiple criteria are defined (see Fig. 1.3) [3–5].

As can be seen from Fig. 1.3, there are capital investments on one side of scales required for D-actions realization, while on the second side there are benefits gained by the D-actions performed:

1. Operation and maintenance costs reduction (C);
2. Overload reduction (P);
3. New consumers connection (W);
4. Power supply security improvement (R), etc.

Part of this gained benefit has monetary value but there are such benefits which can hardly be evaluated financially. Still to initiate optimization it is required.

The economic estimation of D-actions is a thorny task. The D-actions must be estimated applying the method which is named as economic life cycle concept. Capital contributions efficiency is estimated observing operation, maintenance, and technical costs, caused by emergency situations/faults and others.

When estimating D-action, its criterion calculation must not be based on only 1-year figures; the criterion must reflect the whole economic life-cycle of the object.

Fig. 1.3 Estimation of D-actions

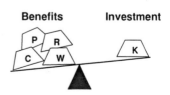

The development optimization is hampered by the fact that it occurs under information uncertainty conditions. It should be taken into consideration in the process of decision-making, and the appropriate methods have to be applied.

Economic criteria. The major economic efficiency ratios are Net Present Value (NPV), Internal Profit Rate (IRR) and Pay Back Period (PB). Net Present Value ratio complies with discounting or monetary value reduction in accordance with discount rate. NPV value is expressed in monetary value units. If NPV > 0, then D-action is economically effective; and the higher NPV, the more effective D-action is. But, if NPV < 0, then D-action is not effective; it is not paid back and causes losses. If NPV $= 0$, then D-action is paid back but is not profitable.

$$\text{NPV} = \sum_{t=0}^{T} C_t \cdot \frac{1}{(1+i)^t} = \sum_{t=0}^{T} C_t \cdot d(t), \tag{1.1}$$

where i income rate, C_t annual costs in development step (D-step) t, $d(t)$ discount rate in D-step t.

In addition to the aforementioned criteria, IRR is used additionally that demonstrates accumulative capacity of D-action (income accumulation speed), and the higher IRR value is, the faster the D-action will be paid back. IRR value is expressed in percentage. According to the situation available in credit resources market and basic principles of LTS investment policy, the D-actions are considered as economically efficient if IRR $> 10\%$.

The payback period of contributions (investments) is a period when the system income sum taking into account monetary value reduction during the period is equal to costs sum of the system. In order to consider the D-action as economically efficient, its payback period shall be shorter than the D-action technical life cycle (depreciation period). The less payback period, the more favorable D-action from economic viewpoint.

In order to estimate these ratios in the preliminary stage of D-actions, cash flow shall be calculated for the whole economical per year life-cycle of respective D-action. If cash flow is calculated including D-action and excluding it, then the ratios of economic efficiency of the D-action can be estimated.

Technical criteria. Technical criteria depend on the specific system. In dynamic models for development optimization, technical parameters limitation is observed not with strict constraint values, but with so-called fuzzy constraint method. Applying fuzzy constraint method the objective function is supplemented by additional criteria (penalty functions) (see Fig. 1.4). Penalty function value within the limits is not high, but as the distances expand, it increases considerably. Applying fuzzy constraint method extra opportunities arise.

Objective function. The major criterion for LTS D-plan is management optimization objective function, which shall display and integrate technical, economic, power supply reliability, ecological, etc., parameters depending on specific technical system [6]. Let us designate this objective function as system D-plan g quality criterion and mark it with $F(T, g)$. Further on, the objective function will be calculated by a formula

Fig. 1.4 Fuzzy constraint method

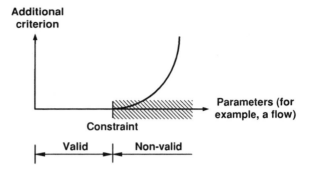

$$F(T,g) = \sum_{t=1}^{T} F(e(t)), \qquad (1.2)$$

where t D-step serial number, T number of D-steps in estimation period, $F(e(t))$ system quality criterion in D-step t and D-state $e(t)$, G D-plan $e(1)$, $e(2)$,…, $e(t)$,…,$e(T)$

According to formula (1.2), the objective function is the sum from $F(e(t))$, i.e., an additive function [7].

Let us assume that a system graph, consumption and generation are constant values at the D-step but D-actions are only realized in transition from D-step $t - 1$ to D-step t. Observing these assumptions, $F(e(t))$ model may look as follows:

$$F(e(t)) = k(t) \cdot \tau(t) \cdot d(t), \qquad (1.3)$$

where $k(t)$ system quality criterion in the first year of the D-step t; $\tau(t)$ number of years in D-step t; $d(t)$ discount (reduction) coefficient in D-step t

Given that the assumed conditions are observed, $F(e(t))$ is not dependent on D-plan up to D-state $e(t)$. Thus, the assumed objective function (1.2) allows dynamic programming application for task solution on LTSs sustainable development management.

1.3 Development Optimization Principles under Uncertainty Conditions

Calculation results credibility depends on initial data credibility and calculation methods.

In the process of systems sustainable development management the following uncertainty groups shall be taken into consideration:

1. Physical uncertainties;
2. Financial uncertainties;
3. Regulators uncertainties.

Selecting optimal D-plans under uncertainty, it is necessary to form: information package set, representing information credibility range (further on package *i*
will be named as prognosis *i*); credibility estimation criteria and comparable
D-plans.

As research data demonstrate, estimating D-plans under uncertainty conditions
is sufficient with relatively small information package set which represents
information credibility range. This is due to the following aspects: objective curve
of optimum in district is very flat and system parameters change discretely.

In LTS sustainable development management, significant information is
demand and prices forecast. In general case, the demand may either increase or
drop. It is essential to improve forecasting methods. However, forecasts will
always be inaccurate. The longer is the forecast period, the higher is prognosis
error. The longer the estimation period, the higher prognosis error.

*Selection of D-plans (decision making) under information uncertainty condi-
tions.* The analysis of LTS sustainable development in compliance with the
methods must be performed for the whole economic life cycle of the object (for
technical system it, in general, is 20–25 years). D-plan estimation period *T* can be
assumed equal or higher than economic life cycle period.

The objective function must be calculated for the whole estimation period
T. Under information uncertainty conditions, the final decision must be made
specifically for not durable time period, not far perspective of 3–5 years. Let
us call this time interval *decision making period* $t_d < T$ and period from t_d up to
T—adaptation period (see Fig. 1.5).

When estimating technical systems sustainable development, *T* must be
assumed longer than economic life cycle period—advisable up to 50 years.

Information on LTS sustainable management estimation during the estimation
period is uncertain, therefore, the decisions made have to be specified regularly.

Development management process is going on gradually; it includes these
stages [8]:

1. Observing external information particularization (specification);
2. Utilizing initial information decisions of previous phases made on specific
   objects (horizontal information flow);
3. Utilizing specified information on previous system development (vertical
   information flow).

Let us review dynamic LTS sustainable development management process
assuming that estimation period is 20 years, which is relatively divided into four
5-year duration phases (see Fig. 1.6). Besides, there is "0"-phase, when technical
economic analysis (feasibility study), decision making and designing of the

Fig. 1.5 Estimation (*T*),
decision making (t_d) and
adaptation (t_{ad}) period

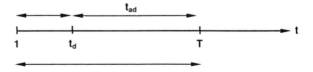

Fig. 1.6 Development management dynamic process [8]

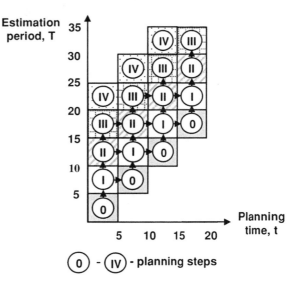

$$\underset{j\in v}{\min} R(j)_{\max} = \underset{j\in v}{\min}\,\underset{i\in \mu}{\max}(F(j,i) - F(i)), \qquad (1.4)$$

selected D-action is performed. The decision is only made on those D-actions that will be realized (constructed or reconstructed) in the Ist phase. In order to perform technical economic analysis with the preliminary approach, it should be assumed what D-actions shall be realized for II, III and IV phases.

The final decisions on these D-actions shall be made during the next phases of LTS sustainable development management (after 5, 10 and 15 years), specifying the respective information [8].

The comparable D-plans are obtained by optimizing the technical system considered and utilizing average data (basic forecast). Because of optimization, up to ten best competitive D-plans are obtained. Further on system analyst, sharing his experience may reduce or supplement this set. The risk analysis is utilized for optimal D-plan selection. Risk analysis enables selection of optimal D-plan under information uncertainty conditions.

For the tasks on large systems sustainable development management the most appropriate is minimal maximal risk:

where $F(i)$ minimal objective function value in case of forecast i, $F(j, i)$ objective function for D-plan j in forecast i, i serial number of forecast, j serial number of D-plan, μ set of forecasts, v set of plans (Tables 1.1, 1.2)

LTS sustainable development adaptation. Decision making and estimation period differentiation ($t_d < T$) demands radical modification of D-plan definition. The respective statement is illustrated in the Table 1.3.

Table 1.1 Risk matrix

	D-plans			
	1	...	j	...
Forecast				
1	$R(1, 1)$	...	$R(1, j)$	...
	...	...	...	...
i	$R(i, 1)$	...	$R(i, j)$	...
...	...	...	...	...

Table 1.2 Risk matrix for wires cross-section selection in the task

	D-plans		
	1 S_1	2 S_2	3 S_3
Forecast			
1	0	0.06	0.25
2	0.10	0	0.15
Max risks	0.10	0.06	0.25

Optimal D-plan is S_2

Table 1.3 Example of LTS sustainable development adaptation

1st forecast			2nd forecast		
t	M_1	M_2	M_1	M_2	
1	A_1	A_2	A_1	A_2	Estimation
2 (t_d)	B_1	B_2	B_1	B_2	Estimation
3	a_1	a_2	a_3	a_4	Adaptation
4 (T)	b_1	b_2	b_3	b_4	Adaptation
Objective function	$F(1, 1)$	$F(1, 2)$	$F(2, 1)$	$F(2, 2)$	

M_1, M_2 D-plans
A, B D-actions, realized in the period of decision-making depend on D-plan specifically but not on forecast
a, b D-actions, realized beyond the period of decision-making, acquired by adaptation (dynamic optimization)

1.4 Problem of Optimal D-Plan Selection

If not only realizable D-actions set is to be optimized but optimal realization time is determined too, then the calculation time required for essentially increasing (1.6). When only one calculation level is considered, optimal realization time usually affects optimal D-plan. Only multi-steps dynamic models ensure adequate model and objective decision motivation. This is due to competitive D-plan relative efficiency dynamic characteristics. Relative efficiency of D-plan g in D-step t can be characterized with D-plan serial number $w(t)$, because in optimization process comparable D-plans are arranged by criterion (1.5) descending values.

$$F(t, g) = \sum_{t=1}^{t} F(e(t)). \tag{1.5}$$

Fig. 1.7 Competitive
D-plans relative efficiency
dynamics

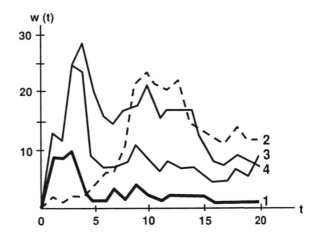

Criterion (1.5) of the D-plan with $w(t) = 1$ has maximal value.

As shown in Fig. 1.7, there are different efficient D-plans in different time periods. D-plan *1*, which has the least objective function, is the less favorable in initial period than the worst D-plan 2. The example is real—it considers transmission network is considered. The task on voltage selection is solved (330 or 750 kV) for the lines transporting power and energy generated by big power plants. In the first D-plan 750 kV is selected, in the second—330 kV.

The reviewed example demonstrates that D-plans are to be compared by objective function value (1.2) during the estimation period. Alternative D-plans will be analyzed to the end of estimation period to make a decision on optimal D-plan.

Figure 1.8 indicates that the number of LTS D-plans is increasing rapidly if D-actions are enlarged and the number of estimation period steps is enlarged too.

Let us calculate total number of LTS D-plans. Each D-plan is characterized by realized D-plans number n and its realization moment T, as well as by each D-action realization type. The total number of D-plans may be calculated by the following formula:

$$V = \prod_{i=1}^{n} V_i = (T+1)^n \prod_{i=1}^{n} m_i, \qquad (1.6)$$

where n number of D-actions, m_i D-action realization type (for example, generation unit capacity) number, i D-action serial number, T number of D-steps in estimation period

If for all D-actions $m_i = 1$, then $V = (T+1)^n$. It means that D-action may be realized in any D-step or not to realize at all.

As can be seen from Fig. 1.8, also in real tasks (for example, $n = 11$, $T = 15$, zone *e*) the number of comparable D-plans attains astronomic quantity (10^{15}). Hence, it is required to apply specialized dynamic optimization methods need to be applied in LTS sustainable development management process.

Fig. 1.8 D-actions number n and number of D-steps T value zones, where (assuming that $m_i = 1$) D-plan index is: a from 0 up to 103; b from 103 up to106; c from 106 up to 109; d from 109 up to 1012; e from 1012 up to 1015; f from 1015 up to 1018; g from 1018 up to 1021; h more than 1021

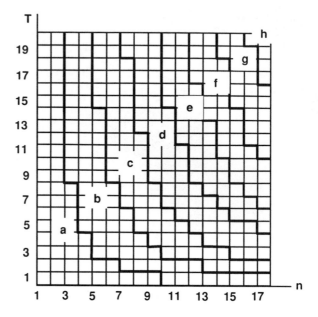

References

1. Riggs JL (1970) Production systems: planning, analysis, control. Wiley, New York
2. Meister D (1976) Behavioral foundations of system development. USA Navy Personnel Research and Development Center. Wiley, New York
3. Neimane V, Anderson G, Krishans Z (1997) Planning of distribution system under uncertainty. In: 32nd Universities power engineering conference UPEC'97, Manchester, UK, Conference Proceedings, pp 1089–1092
4. Cheong HK, Dillon TS (1978) Application of multi-objective optimization methods to the problems of generation expansion planning. In: 6th power systems computation conference, Guilford, Darmstadt, conference proceedings, vol 1, pp 3–11
5. Kersten GE, Mikolajuk Z, Gar-On Yeh A (2000) Decision support systems for sustainable development: a resource book of methods and applications. Kluwer, Dordrecht
6. Hadley G (1964) Nonlinear and dynamic programming. Addison-Wesley, Reading
7. Bellman RE, Dreyfus S (1962) Applied dynamic programming. Princeton University Press, Princeton
8. Boardman JT, Wilde RJ, Richards I, Francis GF (1977) Computer-aided design systems for electricity-network planning. IEEE Power Eng. Soc. Text A. Paper on winter meet (144.9/1–144.9/9). New York

Chapter 2
Mathematical Basics of LTS Sustainable Development Dynamic Management

Abstract This chapter describes the basic concepts of mathematical basis for LTS sustainable development management: LTS development actions, LTS development state set, and LTS development process. Using these terms, mathematical model of development process was elaborated, and algorithmic aspects of development process management were analyzed. This chapter presents the concepts of *development process management recursion equation and optimal initial state*. The analysis of the model has shown that it is possible to calculate LTS optimal development plan considering only OIS set.

2.1 Principal Concepts

Development period of LTS may be represented as a multi-step D-process. To elaborate mathematical basics of dynamic management methods, we use dynamic programming created by Bellman [1, 2] as a starting point, with whose help optimal multi-step tasks solutions may be determined. Dynamic programming is based on the following optimum characteristics: technical system optimal operation depends only on the system D-state at the specific time moment but not on its previous D-plan until the specific moment. This kind of assumption may be applied if optimization criterion is expressed as a sum. Many LTS development management criteria are represented by sum, for example, the total discounted cost criterion.

This chapter deals with selecting dynamic programming recursive equation for LTS development optimization.

In the book, we apply graph and set theory concepts and methods for modeling LTS D-plan and development states (D-states) [5].

Graph theory investigates graph characteristics and related interconnections, as well as elaborates algorithms for its characteristics identification and sub-graphs determination. Initial systematic researches in this area belong to L. Euler. As an independent subject, graph theory was established in 1936 when the world's first book on graph theory written by Hungarian mathematician D. König was published.

Graph is a mathematical model showing interconnections among objects. Graph is described with the following sets: nodes or vertex set V and edge set E. If all the edges are various, that is a unigraph, if edges are repeated—a multigraph. The significant characteristic is node incidence level. Graphs can be guided (see Fig. 2.1) and non-guided. Special graphs are loop, tree, path and planar graphs (see Fig. 2.1).

Set theory is the branch of mathematics that studies sets, which are collections of objects. Although any type of object can be collected into a set, set theory is applied most often to objects that are relevant to mathematics. Set theory, however, was founded by a single paper in 1874 by Georg Cantor: "On a Characteristic Property of All Real Algebraic Numbers".

Set theory begins with a fundamental binary relation between an object o and a set A. If o is a member (or element) of A, we write o ∈ A. Since sets are objects, the membership relation can relate sets as well. A derived binary relation between two sets is the subset relation, also called set inclusion. If all the members of set A are also members of set B, then A is a subset of B, denoted A ⊆ B. For example, {1, 2} is a subset of {1, 2, 3}, but {1, 4} is not. From this definition, it is clear that a set is a subset of itself; in cases where one wishes to avoid this, the term proper subset is defined to exclude this possibility. Just as arithmetic features binary operations on numbers, set theory features binary operations on sets. These are union, intersection, complement, symmetric difference, Cartesian product, and power set.

The mathematical logic [4] as well as combinatorial [3, 6] concepts and methods are also used for system D-plan and D-states creation and analysis provided in the book.

LTS D-plan may be described with the following basic concepts.

LTS D-actions. Basic concept definition *D-actions* is a specific feature of LTS development optimization model; it is not applied in classical dynamic programming model where only concept definitions *D-state* and *optimization steps* are used. We consider that the application of *D-actions* provides opportunity to solve real systems development optimization tasks.

D-actions form development model. Development model is used to determine optimal plan with dynamic optimization model as well as by applying plans

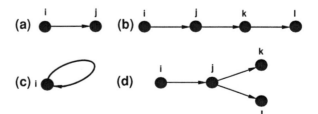

Fig. 2.1 Illustration of graph theory basic characteristics: **a** Guided graph, **b** path, **c** loop, **d** tree

Fig. 2.2 D-states formation
scheme

Fig. 2.3 Formation of information system calculation for
D-state *e(t)*

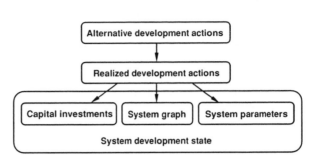

estimation method. Important concept related to system economic analysis complex is possible (alternative) actions.

Let us define D-action. Generally, D-action reflects activity, new system object construction, as well as existing objects extension, reconstruction, and modernization.

D-action is assigned along with capital investments and relevant network elements (elements that are included or excluded from the estimation calculation scheme because of the D-action). The D-actions are assigned by user according to the specific task, experience, etc. Dynamic optimization of user software assigned D-actions is employed as "bricks" for creating D-plan and synthesizing optimal plan. D-actions are systems development dynamic optimization task options. All possible D-actions set constitute this task optimization area. "D-actions" may be interpreted as "purchases". D-action is always interrelated with activity. Such variables differ from traditional approaches in many other optimization models where elements parameters are modeled: we are searching for "what to do" and "how to do"; elements parameters are created because of activities.

D-actions are structured as follows: elementary and compound. The compounds are formed from elementary actions, applying *logical conditions*: "one from" and "after".

D-states formation scheme is depicted in Fig. 2.2, but information scheme of D-state *e(t)* calculation is shown in Fig. 2.3.

Definition *D-action realization* denotes respective object construction, reconstruction or liquidation completion (commissioning or decommissioning), but not time period of object construction, reconstruction, or liquidation. It is assumed in the model that D-action is realized at the beginning of the first year of D-step. Constructions, reconstructions, or dismantling are finished within this year. This time interval duration is observed when the respective D-action costs (capital investments) are calculated. As soon as the D-action is realized, estimation calculation scheme is altered.

Possible realization period. The beginning of this period is usually determined by time length required for object construction, reconstruction, or liquidation. Usually, its completion is concurrent to estimation period completion, but under specific conditions, eventual realization period may be finished faster, if, for example, on plot where construction of new substation is envisaged, other civil works might be started prior to calculation period completion.

D-actions will be marked with $x(i)$, number of D-actions—with n.

LTS D-state. Technical system D-state in optimization task is specified by system graph, its elements parameters and other technical, economic, ecological, etc. parameters. Let us mark D-state with $e(j)$, where j is a sign or sign group, which determines D-state reciprocal dissimilarities. As a distinctive sign, D-step serial number t can be used. Then D-plan is

$$g(k) = e(0), e(1), e(2), \ldots, e(t), \ldots, e(T). \tag{2.1}$$

D-state $e(0)$ is existing D-state, where no any D-action is realized. In common, case D-state may be defined as realized D-actions within the period 0–t:

$$e(t) = e(0) \cup \{x(1)\} \cup \{x(2)\} \ldots \cup \{x(t)\}, \tag{2.2}$$

where $\{x(t)\}$—all step t initial realized D-actions set.

D-state can be classified as technically valid or technically non-valid—if at least one technical limit in D-state is not observed.

LTS D-step. LTS D-step is time period when none of D-actions is realized—system graph and its parameters are not changed. In D-step period, it is assumed that annual consumption of the first year of D-step, 24 h, seasonal and monthly consumption is preserved for the period of all further years. Performing mathematical calculations, we assume that all D-step period realized D-actions are realized simultaneously, going on from D-step $t - 1$ to D-step t. In such a way, real continuous calculation process is modeled by discrete D-plan. In general, D-step length is from 1–3 up to 10–13 years. In the initial phase of estimation period D-steps are shorter, the next D-steps length is longer, but the last D-steps are the longest.

LTS D-plan. In estimation period, which is divided into T D-steps, may be various D-states $e(t)$. A sequence of D-states $e(t)$, if $t = 1, 2, \ldots, t, \ldots, T$, is called large technical system D-plan, which will be marked with $g(k)$. Graphical model of D-plan is guided path graph (see Fig. 2.4).

Graph vertices and edges can be logical admissible or non-admissible (if at least one logical condition in D-plan is not observed).

D-plan $g(k)$ may be unambiguously characterized by realized D-actions and their realization time as well as estimated by summary, at the initial moment reduced, system quality criterion—objective function (investments, generation,

Fig. 2.4 Large technical system D-plan

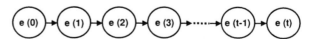

consumption, generating unit production range, transport network flows and overloads).

D-plans are classified into two groups:

1. Logically admissible D-plans—there is no D-state where all logical conditions are observed;
2. Logically non-admissible D-plans—in any graph vertex or edge at least one logical condition is not observed; in order to operate non-admissible D-states, it is required to realize more supplementary D-actions, i.e. to create another D-plan $g(k)$.

LTS D-states set. In optimization process, LTSs are required to be reviewed not only as separate D-states but also as a set of different D-states. Let us mark D-states set with E. To distinguish different types of D-states, let us use symbol $E(i)$. The major types of D-states are the following:

1. D-states set $E(t)$, which integrates all of D-step t possible D-states;
2. Subsets of set $E(t)$: (a) D-states subset, in which D-states are integrated forming development sequel process, preserving the existing D-states or realizing additional D-actions; such subset is marked with $E[t, e(i) \rightarrow]$ and (b) D-states subset $E[t, e(i) \leftarrow] \subset E(t)$, in which D-states are integrated in D-step $t - 1$, from which by realizing one or several D-actions (or not realizing any), D-state $e(i)$ can be formed. It must be taken into consideration that in specific D-state $e(i)$ only one D-action $x(j)$ is allowed, because D-action is the realized fact and cannot be cancelled;
3. Subsets of set $E(t)$ that integrate D-states with equal number of realized D-actions.

Finally, let us review *operations with D-states sets* used in dynamic optimization process:

1. Membership of D-state e in set E:$e \in E$;
2. Calculation of function $F(e)$ for all $e \in E$: $F(e)$;
3. Union of two D-states sets $E(1)$ and $E(2)$: $E(3) = [E(1) \cup E(2)]$.

2.2 Graphical Model of Development Process

LTS D-states $e(i)$ can be regarded as graph G vertices, but graph edges are lines that connect two D-states (graph G vertices). Graph edges represent probable transition from D-state $e(t - 1)$ to D-state $e(t)$. Taking into consideration that LTS D-state is a set of realized D-actions, $e(t) = \{x\}$ $[e(0) = es]$ (where es—empty set), the transition from $e(t - 1)$ to $e(t)$ is possible if $e(t - 1) \subset e(t)$. Thus, graph G edge corresponds to realized D-actions set $\{x\}_p$, within transition time; in particular cases there is only one D-action or that is also an empty set $\{x\} = es$. For Graph G edges there is only one direction—from $e(t - 1)$ to $e(t)$.

Fig. 2.5 LTS graph; existing elements are represented by *uninterrupted line*, but new constructed system elements—by *interrupted line*

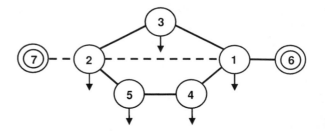

Figure 2.5 shows a system with five consumption nodes and one generation node. The highest consumption is in node 2, which is located within the furthest distance from generation node. Due to consumption rise in transport network, links 1-3-2 and 1-4-5-2 loads are close to the valid range. In addition, generation node is almost fully loaded. In order to prevent critical situation in the system, four alternative D-actions may be applied:

1. Reconstruction of existing links 1-3-2;
2. Reconstruction of existing links 1-4-5-2;
3. Construction of new links 1-2;
4. Construction of new generation node 7 at a higher load of node 2.

D-action combination set is given in Table 2.1.

Let us mark various D-states with $e(t, \chi)$. Development graph of technical system considered (if $T = 5$) is depicted in Fig. 2.6.

System development by T D-steps is illustrated with a succession of D-states (graph G vertices connected with graph edges), that is with graph loop g, which starts in vertex $e(0, 0)$ and ends in one of vertices $e \in E(T)$. For example, one of admissible graph loops is a line, which connects graph vertices sets $e(1, 3)$, $e(2, 3)$, $e(3, 3)$, $e(4, 6)$ and $e(5, 6)$. For this D-plan summary reduced system quality criterion can be calculated for five D-steps $F(5, g)$.

If system graph G is given, then all possible system D-plans are given too.

System development graph vertices set. Let us consider these sets:

1. Vertices set in D-step $t - E(t)$;
2. Vertices set in D-step t subsets, in which all vertices have characteristics $i - E(t, i)$;
3. D-step t vertices subset $E[t, E(t - 1, i) \rightarrow]$, where all vertices are related to D-step $t - 1$ vertices $e(t - 1) \in E[(t - 1, i)]$ set $E[t, E(t - 1, i) \rightarrow]$, we mark the vertices that are related to sub-set $E(t - 1, i)$. In special case, a set may contain a single vertex, then vertices $e(t) \in E(t - 1, i)$ are interrelated only with vertex $E(t - 1, i)$;
4. Initial vertices set $E(t - 1, i)$ related to sub-set $E(t, i)$.

Let us explain definitions of the subsets considered with four examples which are illustrated in Fig. 2.3.

Table 2.1 LTS D-actions sets

No.	Code χ	Set file
1	0	Empty set
2	1	1
3	2	2
4	3	3
5	4	4
6	5	3; 1
7	6	3; 2
8	7	4; 1
9	8	4; 2

Fig. 2.6 Technical system development graph

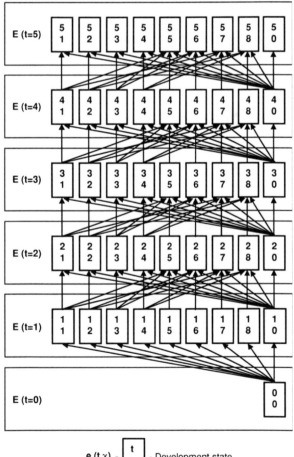

Example 1. A vertex set with a single vertex $e(2, 3)$ is given. Then with the given set interrelated vertices set is $\{e(3, 3), e(3, 7), e(3, 8)\}$.

Example 2. A vertex set $\{e(2, 4)\}$ is given. Then with the given set related vertices (development follow-up) set is $\{e(3, 4), e(3, 6), e(3, 7), e(3, 10)\}$.

Example 3. A vertex set $E(t - 1, i) = \{e(2, 2), e(2, 6)\}$ is given. Then with the given set related vertices set is $\{e(3, 2), e(3, 6), e(3, 8)\}$.

Example 4. A vertex set $E(t - 1, i) = \{e(3, 3), e(3, 5)\}$ is given. Then initial D-states set is $E[(t - 1, i) \rightarrow] = \{e(2, 3), 2(2, 5), e(2, 0)\}$.

The following conclusions result from the reviewed concept definitions:

1. In all D-steps, all D-states and transitions must be preserved. Otherwise, D-plan is terminated.
2. Usually, the reason why D-states are not preserved is that technical constraints are not met. Generally, irrelevant D-states are occurring by incomplete D-states set.
3. In order to prevent the occurrence of irrelevant D-states, it is not allowed to utilize rigid constraint method; instead, fuzzy constraint method and penalty function must be applied. Then the first optimization can be finalized. After optimization, the results can be analyzed and D-action set can be supplemented.

2.3 Mathematical Formulation of Development Process Optimization Task

The target of LTS development optimization is to determine and identify an admissible D-plan, in which the maximal reduced system development quality criterion—objective function would be met in T D-steps. Such D-plan will be marked with optg $[T, E(T)]$ and respective objective function will look like this

$$f[T, E(T)] = \max_{g \in G} F(g). \tag{2.3}$$

The right-hand side of expression 2.3 shows that objective function maximization is taking place (if objective function incorporates also income from the sold production) over all graph G loops (routes). It may seem that the most elementary optimization is to consider in succession all probable graph loops. Still this kind of method is only admissible when D-plans scope is not considerable; but as shown in Fig. 1.8, in real tasks, astronomic figures may be obtained and calculation of D-plans might take several years. Optimization method that does not cover analysis of all D-plans $g \in G$ would be more efficient.

Let us assume that in given D-step t information available is:

1. Valid D-states set, $e(t)$ for step t;
2. Objective function $F[e(t)]$ for development from $e(0)$ to $e(t) \in E(t)$;

3. Valid D-states set, $E(t - 1)$ for step $t - 1$;
4. D-states set from which transition is probable to $e(t, i)$:

$$E[t, e(t - 1) \to]. \tag{2.4}$$

The transitions from step $t - 1$ D-states to step t D-states $e(t)$ are determined for all $e(t - 1) \in E(t - 1)$ or all graph edges are given that connect $e(t - 1)$ with $e(t)$, as well as respective D-actions characteristics k and system graph changes;

5. Step $t - 1$ D-states set from which the given D-state $e(t)$ may be attained;

$$E[(t - 1), e(t) \leftarrow]. \tag{2.5}$$

Sets 2.5 are given for all D-states $e(t) \in E(t)$.

Besides, let us assume that for all $e(t - 1) \in E(t - 1)$ there is a certain optimal D-plan optg $[t - 1, e(t - 1)]$.

Let us mark with $f[t, e(t)]$ the maximal summary reduced criterion value within the period up to D-step t and D-state $e(t)$, i.e. is for optimal development optg $[t, e(t)]$. Let us call this value functional. It required to calculate for all D-states $e(t) \in E(t)$, if values $f[t - 1, e(t - 1)]$ are known for all D-states $e(t - 1) \in E(t - 1)$.

Functional $f[t, e(t)]$ must be expressed observing income $(+)$ from production sold and expenditures $(-)$:

$$f[t, E(t)] = \max_{e(t) \in E(t)} f[t, e(t)]. \tag{2.6}$$

Formula of functional calculation may be expressed as follows:

$$f[t, e(t)] = F(e(t)) + \max_{e(t-1) \in E[t-1, e(t) \leftarrow]} f[t - 1, e(t - 1)], \tag{2.7}$$

where $F(e(t))$—reduced system quality criterion in D-step t and D-state.

Formula 2.7 is dynamic programming recursive equation that is used to optimize LTS D-plan.

In the first D-step of estimation period functional $f[t - 1, e(t - 1)]$ is system summary reduced quality criterion for previous D-steps that is independent of system further development and may be omitted in calculation. Assuming that $f[0, e(1)] = 0$, we have

$$f[t - 1, e(1)] = F(e(1)). \tag{2.8}$$

Using recursive formula 2.7, starting from the first D-step $t = 1, 2, 3,...,T$, in inductive way, one may calculate functional $f[t, e(t)]$ values for all D-steps and D-states and determine optimal D-plan $f[t, e(t)]$.

2.4 Algorithmization Aspects of Development Management Process

LTS development optimization applying dynamic programming recursive Eq. 2.7 can be performed in different ways depending on how recursive Eq. 2.7 is written down.

First, let us review the option in writing:

$$f[t, e(t)] = F[e(t)] + \max_{e(t-1) \in E[t-1, e(t) \leftarrow]} f[t-1, e(t-1)]. \qquad (2.9)$$

Recursive Eq. 2.9 is used if all D-states are considered $e(t) \in E(t)$ for each D-state $e(t)$:

1. To calculate step t objective function item $F[e(t)]$—to perform system D-state technical, economic and ecological criteria calculation.
2. To define D-states set $E[t-1, e(t) \leftarrow]$, from which integrated D-states $e(t-1)$ may reach D-state $e(t)$.
3. To calculate the set functional maximal value $\max_{e(t-1) \in E[t-1, e(t) \leftarrow]} f[t-1, e(t-1)]$.

Thus, optimal initial state is defined $e(t-1)$ and optimal development up to D-state $e(t) - optg \{t-1, E[t-1, e(t) \leftarrow]\}$.

In order to realize this algorithm, it is required to form operative memory data array $f[t-1, e(t-1)]$ for all D-states $e(t-1) \in E(t-1)$ and $f[t, e(t)]$ all D-states $e(t) \in E(t)$. Comparing with all D-plans calculation, the algorithm observed essentially reduces calculation time. Still, capacious computer memory consumption is required for storage of calculation interim results. It considerably limits capabilities of method application.

Let us review modifications of such recursive equation which enable are to use calculation algorithm without saving in data array functional $f[t-1, e(t-1)]$ and $f[t, e(t)]$ values on all D-states sets $E(t-1)$ and $E(t)$.

In order to elaborate such algorithm, the answer to the question must be known. By what indications the D-states, which can be used in optimal D-plan, must be identified? If such indications are available, there is no need to bear in mind functional values for all D-states.

Let us mark with $\Omega(t-1) \subset E(t-1)$ D-states $e(t-1)$ set that contains step t optimal initial states.

D-states that belong to set $\Omega(t-1)$, will be marked with $e(t-1, \omega) \in \Omega(t-1)$, where $\omega = 1, 2, 3,...,\omega(max)$.

The set $\Omega(t)$ may be defined by the following method (see Fig. 2.7). D-state $e(t-1, \omega = 1)$ is determined by formula

$$f[t-1, e(t-1, \omega = 1)] = \max_{e(t-1) \in E(t-1, \omega=1)} f[t-1, e(t-1)], \qquad (2.10)$$

where $E(t-1, \omega = 1) = E(t-1)$.

Fig. 2.7 D-states sets Ω $(t-1)$ (required for optimal development estimation/ calculation) elaboration algorithm illustration

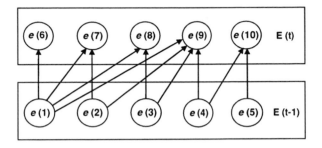

$$E(t-1, \omega=1) = \{e(1), e(2), e(3), e(4), e(5)\}; \quad e(t-1, \omega=1) = e(1);$$

$$\Omega(t-1, \omega=1) = \{e(t-1, \omega=1), \ldots\};$$

$$E[t, \Omega(t-1, \omega=1) \rightarrow] = \{e(6), e(7), e(8), e(9)\};$$

$$\overline{\Omega}(t-1, \omega=1) = \{e(1), e(2), e(3)\}; \quad \overline{\overline{\Omega}}(t-1, \omega=1) = \{e(1), e(2), e(3)\};$$

$$E(t-1, \omega=2) = \{e(4), e(5)\}; \quad e(t-1, \omega=2) = e(4);$$

$$\Omega(t-1, \omega=1) = \{e(t-1, \omega=1), e(t-1, \omega=2)\};$$

$$E[t, \Omega(t-1, \omega=2) \rightarrow] = \{e(6), e(7), e(8), e(9), e(10)\};$$

$$\overline{\Omega}(t-1, \omega=2) = \{e(1), e(2), e(3), e(4), e(5)\};$$

$$\overline{\overline{\Omega}}(t-1, \omega=2) = \{e(1), e(2), e(3), e(4), e(5)\}; \quad \Omega(t-1) = \{e(1), e(4)\}.$$

D-state $e(t-1, \omega=1)$ is optimal development by $t-1$ steps final state.

For D-state $e(t-1, \omega=1)$, related set is defined as

$$E[t-1, e(t-1, \omega=1) \rightarrow]. \tag{2.11}$$

Sets 2.11 initial set is the set

$$\text{initial } E[t-1, e(t-1, \omega=1) \rightarrow]. \tag{2.12}$$

D-states set 2.12 comprises D-state $e(t-1) \in \Omega(t-1, \omega=1)$, from which transition is only possible to set 2.11.

In general case, the transition from set $\Omega(t-1, \omega=1)$ to set $E(t)$ is possible in many ways, i.e. to several D-states $e(t)$. Each D-state $e(t)$ has its own system quality criterion value:

$$F[t, e(t)] = f[t-1, e(t-1)] + F[e(t)]. \tag{2.13}$$

Formula 2.13 shows that optimal development may not occur through D-state $e(t-1) \in \Omega (t-1, \omega=1)$, for which the following condition is valid

$$f(t-1, e(t-1)) + F(e(t)) < f(t-1, e(t-1, \omega=1)) + F(e(t)). \tag{2.14}$$

This condition can be verified even without calculating objective function component $F(e(t))$ of step t for D-states $e(t)$. Excluding from set $\Omega(t-1, \omega=1)$ all D-states that do not conform to condition 2.14, we obtain D-states set $\Omega(t-1,$

$\omega = 1$), in which optimal development can be determined only from D-state $e(t - 1, \omega = 1)$.

It often happens that condition 2.14 is valid for all D-states $e(t - 1) \in \Omega(t - 1, \omega = 1)$. In such cases $\overline{\overline{\Omega}}(t - 1, \omega) = \overline{\Omega}(t - 1, \omega)$.

D-states $e(t - 1, \omega = 1)$ are calculated by formula:

$$f[t - 1, e(t - 1, \omega = 2)] = \max_{e(t-1)\in E(t-1,\omega=2)} f[t - 1, e(t - 1)], \qquad (2.15)$$

where

$$E(t - 1, \omega = 2) = E(t - 1) - \overline{\overline{\Omega}}(t - 1, e(t - 1)). \qquad (2.16)$$

Let us mark with $\Omega(t - 1, \omega)$ D-states set

$$\Omega(t - 1, \omega) = \{e(t - 1, \omega - 1), e(t - 1, \omega - 2), \ldots, e(t - 1, \omega)\}. \qquad (2.17)$$

The related D-states set we express as

$$\overline{\Omega}(t - 1, \omega = 2) = \overline{\Omega}(t - 1, \omega = 1) \cup OIS\ E[(t, e(t - 1, \omega = 2)) \rightarrow]. \qquad (2.18)$$

From set 2.18, we exclude all D-states which do not meet this condition:

$$f[t - 1, e(t - 1)] + F(e(t)) < f[t - 1, e(t - 1, \omega)] + F(e(t)). \qquad (2.19)$$

The reviewed analysis allows us to draw the conclusion: technical system development graph loop that is relevant to optimal D-plan may comprise only vertices $E(t - 1)$, which are included in the subset of optimal initial states set $\Omega(t - 1)$:

$$e(t - 1, \omega) \in \Omega(t - 1). \qquad (2.20)$$

D-state $e(t - 1, \omega)$ can be defined using formula:

$$f[t - 1, e(t - 1, \omega)] = \max_{e(t-1)\in E(t-1,\omega)} f[t - 1, e(t - 1)], \qquad (2.21)$$

where

$$E(t - 1, \omega) = E(t - 1) - \overline{\overline{\Omega}}(t - 1, \omega - 1). \qquad (2.22)$$

In order to calculate functional $f[t, E(t)]$ and large technical system optimal development up to any D-step t, it is required in D-step $t - 1$ to keep data on all $[\omega(\max)]$ optimal initial states $e(t - 1)$.

D-states set E is set's $E(t, \omega)$ subset, where is D-state $e(t, \omega)$:

$$E = E(t, \omega) \cap E[t, e(t - 1, \omega)]. \qquad (2.23)$$

Functional $f[t - 1, e(t - 1, \omega)]$ value can be calculated using the following recursive equation:

$$f[t, E(t, \omega)] = \max_{e(t-1,\omega)\in\Omega(t-1)} f[t - 1, e(t - 1, \omega)] + \max_{e(t)\in E} F(e(t)), \qquad (2.24)$$

where

$$E(t, \omega) = E(t) - \overline{\overline{\Omega}}(t, \omega - 1). \qquad (2.25)$$

Recursive Eq. 2.24 may be used to calculate functional value for the given D-states set $E(t, \omega)$ with the algorithm described below.

1. Review all D-states $e(t - 1, \omega) \in \Omega (t - 1)$ from $\omega = 1$ up to $\omega = \omega(\max)$.
2. For each D-state $e(t - 1, \omega)$ define $\max_{e(t)\in E} F[e(t)]$. This task corresponds to LTS statistical optimization, if optimization area is D-states set E. The task solution is D-state with calculable $F[e(t)]$ maximal value. The D-state must conform to these two conditions:

   - D-state is relevant to set $E(t, \omega)$,
   - The transition from D-state $e(t - 1, \omega)$ to $e(t)$ must be possible.

3. From all sums $\max \sum_{e(t-1,\omega)\in\Omega(t-1)} (F(e(t)) + f[t - 1, e(t - 1), \omega])$ the maximal sum shall be calculated, considering all sets $\Omega(t - 1)$ D-states.

Simultaneously applying formulas 2.24 and 2.25, functionals $f[t, E(t, \omega)]$ must be calculated for all ω values $\omega = 1,2,...,\omega(\max)$.

Among D-states $e(t, \omega) \in \Omega (t)$ there is also an optimal D-plan from $t = 1$ up to $t = T$ (in estimation period) D-state $e(t, \omega)$ in D-step t.

LTS optimization task solution is D-state $E(T, \omega = 1)$ with summary reduced system quality criterion $f[T, E(t, \omega = 1)]$.

Comparing both considered quality optimization algorithms; the following conclusions can be drawn:

1. The application of recursive Eq. 2.9 makes algorithms execution more simple. Algorithm drawback is the demand for all potential D-states $E(t - 1)$ and $E(t)$ to keep calculation interim results and to calculate system quality criteria. D-states number in set $E(t)$ is

$$V_E = 2^n, \qquad (2.26)$$

where n—D-actions number. If $n = 20$, which is possible in real optimization tasks, then $V_E = 2^{20} = 10^6$.

2. Applying recursive Eq. 2.24 and expression 2.25, it is not required to keep calculation interim results for all D-states $E(t - 1)$ and $E(t)$. This algorithm allows one to simultaneously utilize dynamic optimization methods, applying summary reduced system quality criterion, and system D-state static optimization to one consumption level (one D-step). If static optimization methods of LTSD-states are effective, then recursive Eq. 2.24 would be effective.

Unfortunately, such methods have not been established yet and therefore the authors are not familiar with them.

References

1. Bellman RE (1957) Dynamic programming. Princeton University Press, Princeton, New Jersey
2. Bellman RE (1961) Adaptive control processes: A. Guided Tour. Princeton University Press, Princeton, New Jersey
3. Hall MJR. (1967) Combinatorial theory. Blaisdell Publishing Company, Waltham, Massachusetts-Toronto-London
4. Mendelson E (1997) Introduction to mathematical logic (4th ed.), Chapman & Hall, London, ISBN 978-0-412-80830-2
5. Ore O (1962) Theory of graphs. American mathematical society, Colloquium Publications, Volume XXXVIII, Providence, Rhode Island
6. Riordan J (1958) An introduction to combinatorial analysis. John Wiley & Sons, Inc., New York, Chapman & Hall, Limited, London

Chapter 3
Modeling LTS and their D-Process

Abstract In LTS development process modeling, alternative links and development actions, as well as action logical conditions are used. This method gives opportunity to observe real wide development task set and to create simple algorithms. To ensure short calculation time it is proposed to express development state as binary vector and to apply mathematical logic operations. It gives opportunity to create LTS development process formation methods, which include simple quick-operating algorithms. This chapter describes static system optimization method for optimal initial state calculation. PSMM research work has shown that such methods have not to the best expected results, therefore, Laboratory PSMM made another choice: research of technical system optimal development regularization and opportunity to apply gradient methods to optimal initial state searching. This approach gave beneficial results—new original dynamic optimization methods group "optimal initial states method" has been elaborated. This method is presented in Chaps. 5 and 6 of the book. Dale et al. (Dynamic Programming in Calculation of Power Networks Development 1969); Dale et al. (Dynamic Optimisation of Electric Powder Networks Development 1990); (Krishans Изв. АН СССР, Энергетика и транспорт 5:32–41, 1981); Krishans et al. (13th Power Systems Computation Conference 2:863–869, 1999); Krishans et al. (Software for Power Supply Utilities Management 2001); Krishans et al. (7th WSEAS/IASME International conference on Electric Power System 2007).

3.1 LTS Modeling Principles in Development Management Process

A mathematical model of LTS is the system and it D-process, which makes it possible to calculate and assess technical, economic and ecological criteria of the analyzed system with a view to make justified decisions on system sustainable development management.

Z. Krishans et al., *Dynamic Management of Sustainable Development*,
DOI: 10.1007/978-0-85729-062-5_3, © Springer-Verlag London Limited 2011

To enable computer implementation of the given task, data are required as well as appropriate software providing optimization algorithm and algorithms for calculating generating nodes output and transport network links load, which is the basis of system quality criteria estimation. Taking into consideration calculation size, methods to be applied must be fast. Similar requirements are also imposed to data.

The input data may be classified into four groups:

- Group 1 data unchangeable within the whole calculation process;
- Group 2 data specific for each D-step;
- Group 3 data specific for each D-state;
- Group 4 data used for analysis of results.

Let us review all groups in detail, assuming technological systems at the generalized level. Each technical system is specific with its particular data (for example, electric power engineering, gas, oil refinery, and industrial concern) which are reviewed in the book as an example only for electric power engineering system (Chaps. 8–10).

Group 1   In general, LTS may be modeled by a graph. This issue has been partly considered in Chap. 1 (Sect. 1.1, Fig. 1.1) and Chap. 2 (Sect. 2.2, Fig. 2.2). In this chapter let us use the system graph shown in Fig. 3.1. When compared with Fig. 2.2, the number of D-actions has increased from 4 to 10.

The following information is obligatory on the graph links:

lc   Link code;
$n(1)$ Initial node;
$n(2)$ End node;
α Link existence mark: $\alpha = 1$—link is in the scheme, $\alpha = 0$—link is not in the scheme, non-realized D-action, interrupted line;
ls   Link length;
Ps   Link capacity;
jsp   Link parameters code that provides opportunity to identify values required for system quality criterion calculation, including production capacity flow per link and its transport costs calculation.

The following information on Graph nodes is obligatory:

Fig. 3.1 LTS graph

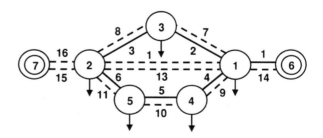

For consumption nodes:

nc   Node code;
P   Consumption in D-step, operational state;

For generation nodes:

ig   Node code;
P1   Maximal generation capacity;
Cg   Product selling price.

The software must provide a procedure for calculating optimal generation capacity in D-step and D-state.

Group 2 Depending on D-step, the following information must be put in the forward to computer: to consumption node: consumption prognosis and product selling price; to generation node: consumption prognosis and product selling price. These data must be sent once in each D-step.

Group 3 Data that are specific for each D-state represent information on realized D-actions set. The creation of system graph must be fast with the least computer run-time consumed.

Group 4 These are: data that provide D-plan determination from the last D-step T in reverse direction and data required for risk analysis of information in uncertainty conditions—these are the main data about development conditions forecast.

3.2 Modeling Methods of LTS D-State in Development Optimization Process

LTS development optimization is aimed to determine a sequence of D-actions which would provide admissible (with technically valid and logically admissible D-states) development in estimation period with maximal system quality criterion summary reduced value. Such a definition of development optimization is valid for any technical system.

For example, a technical system represented in Fig. 3.1 may raise a variety of issues to be considered in optimizing system development such as: whether it is required to reinforce the system in principle; whether the existing transport network links are to be rebuilt and in what sequence; whether new transport network links 1–2 is to be built and with what capacity (P1 or P2); whether the maximal capacity of existing generation node 6 is to be increased; whether a new generation node 7 is to be constructed and with what maximal capacity.

The purpose of optimization task is to calculate D-action realization time t_r. If $t_r > T$, then there is no need to realize the D-action.

Table 3.1 The list of D-actions

Action no.	Action characteristics	Link code (is)	α_{is}^a	Logical condition
1	Reconstruction of links 1–3–2 increasing transmission capacity	7	+	Action 1 or 2
		2	−	Action 1 or 3
		8	+	
		3	−	
2	Reconstruction of links 1–3 increasing transmission capacity	7	+	Action 2 or 1
		2	−	
3	Reconstruction of links 3–2 increasing transmission capacity	8	+	Action 3 or 1
		3	−	
4	Reconstruction of links 1–4 increasing transmission capacity	9	+	Action 4 or 1
		4	−	
5	Reconstruction of links 4–5 increasing transmission capacity	10	+	
		5	−	
6	Reconstruction of links 5–2 increasing transmission capacity	11	+	
		6	−	
7	Links 1–2 construction with transmission capacity Pl_1	12	+	Action 7 or 8
8	Links 1–2 construction with transmission capacity Pl_2	13	+	Action 8 or 7
9	Reconstruction of generation node 8 increasing capacity	14	+	
		1	−	
10	Construction of new generation node 7 with capacity P_1	15	+	Action 10 or 11
11	Construction of new generation node 7 with capacity P_2	16	+	Action 11 or 10

[a] Link code is connected to system graph (+) or removed from the system graph (−)

Table 3.1 summarizes the D-actions considered. In the optimization task, ten D-actions or 1,024 various D-states have been considered.

The following D-actions are reviewed in the example:

D-action 1—concurrent reconstruction of transport links 1–3–2. D-action system graph is supplemented with two links 7 and 8, but links 2 and 3 are removed. The parameters of new links and removed links differ.

D-actions 2 and 3 provide opportunity to gradually consider (links 1–3 and links 3–2) the reconstruction of links 1–3–2.

D-actions 4, 5 and 6 provide opportunity to consider the reconstruction of the whole links 1–4–5–2 either gradually or simultaneously.

D-actions 7 and 8 are modeling the construction of new links 1–2, considering two link options. In the option with two new links, logical conditions of D-actions are not provided.

D-action 9 is modeling the reconstruction of the existing generation node.

D-actions 10 and 11 are modeling the construction of new generation node 7 in two options—generation capacity P_1 or P_2. The construction of two generation nodes is not acceptable, this condition is modeled by two logical conditions "or".

Fig. 3.2 System graph $e(0)$
transition as system graph in
D-step t for D-state $e(t)$

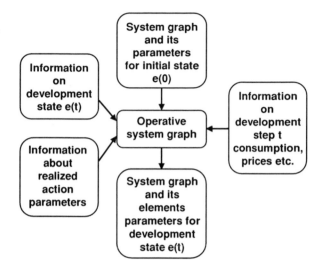

Besides, economic and ecological ratios are given to each D-action: investment;
operation costs of system element (which D-state or parameters are altered when
realizing D-action); amount of harmful emissions etc.

Let us review calculation scheme of total reduced quality criterion for an LTS
in system D-state $e(t)$ (see Fig. 3.2).

The major issue of LTS D-state illustration is representation method of realized
D-actions set. It must meet the following requirements (taking into consideration a
considerable number of D-states):

1. For recording one realized D-actions set, random access memory must be
   utilized;
2. Set formation algorithm must be simple and fast;
3. Set decoding algorithm for system development calculation process must be
   simple and fast;
4. Also verification $e(t - 1) \subset e(t)$ (i.e., transition possibility from D-state in step
   $t - 1$ to D-state in step t) algorithm must be simple and fast.

These requirements may be represented by realized D-actions set in a vector
format:

$$e(t) = \chi = \{M_n, M_{n-1}, \ldots, M_i, \ldots, M_2, M_1\}, \qquad (3.1)$$

where Mi D-action i realization type, i D-action index number, n number of
D-actions.

Let us assume that in the task under consideration, D-actions are allowed to be
used that contain only two D-states: non-realized $= 0$ and realized $= 1$. More
types of realized D-actions are also admissible, for example: construction of new
links 1–2, considering two options; new generation node construction in node 7,
considering two options by generation capacity. As shown in the example

(see Table 3.1), in such cases the number of D-actions must be increased $(7 + 8)$ and $(10 + 11)$, and these logical conditions must be applied: "or", "one from the set", and "after".

If D-action comprises only two values, 0 or 1, then vector (3.1) is binary vector—line that may be recorded in integer type variable extended cell up to $n = 30$.

Let us review two D-states in our example.

D-actions realized in D-state 1:

1. D-action 1 (reconstructed links 1–3–2),
2. D-action 4 (reconstructed links 1–4),
3. D-action 5 (reconstructed links 4–3),
4. D-action 6 (reconstructed links 3–2);
5. D-action 9 (increased capacity of the existing generation node).

D-actions realized *in* D-state 2:

1. D-action 7 (links 1–2 is constructed—option 1),
2. D-action 10 (new generation node is constructed—option 1)

D-states binary vectors are given in Table 3.2.

Consequently, the D-state 1 the binary vector $e(1)$ is expressed by the integer number $\chi(1)_{10} = 413$, but the binary vector $e(2)$ is expressed as $\chi(2)_{10} = 576$. It means that little random access memory is required for D-states storage, even if the number of D-states to be stored is considerable—up to 10,000.

Let us now review a method and algorithms for decoding D-state $\chi(i)$ and forming technical system graph for D-state $e(t, i)$.

The method is based on binary unit diagonal matrix (UDM) (see Table 3.3).

Matrix row is binary unit vector UDM (i) with length n.

Table 3.2 Description of D-states with binary vectors

i	2^{i-1}	M_1	M_2	$M_1 \cdot 2^{i-1}$	$M_2 \cdot 2^{i-1}$
1	1	1	0	1	0
2	2	0	0	0	0
3	4	0	0	0	0
4	8	1	0	8	0
5	16	1	0	16	0
6	32	1	0	32	0
7	64	0	1	0	64
8	128	0	0	0	0
9	256	1	0	256	0
10	512	0	1	0	512
χ				413	576

where i D-state vector element index number, M_1 vector element of D-state 1, M_2 vector element of D-state 2

Table 3.3 Unit diagonal matrix—UDM(i, j)

i	j									
	1	2	3	4	5	6	7	8	9	10
1	1	0	0	0	0	0	0	0	0	0
2	0	1	0	0	0	0	0	0	0	0
3	0	0	1	0	0	0	0	0	0	0
4	0	0	0	1	0	0	0	0	0	0
5	0	0	0	0	1	0	0	0	0	0
6	0	0	0	0	0	1	0	0	0	0
7	0	0	0	0	0	0	1	0	0	0
8	0	0	0	0	0	0	0	1	0	0
9	0	0	0	0	0	0	0	0	1	0
10	0	0	0	0	0	0	0	0	0	1

Fig. 3.3 D-state decoding algorithm

```
begin;
for i:=1 to n do begin;
k:=(χ^UDM[i]);
if k=0 then raχ i):=0
else raχ(i):=1;
end;
end;
```

Table 3.4 Conjunction and disjunction conditions for all D-state vector elements

χ	UDM(i)	Conjunction	Disjunction
1	1	1	1
1	0	0	1
0	1	0	1
0	0	0	0

The decoding of D-state binary vector χ is formed by one-dimension integer type data array raχ(30) (realized D-actions in D-state e). Decoding algorithm is depicted in Fig. 3.3.

where k integer type operable cell, $\wedge$ logical "and" (conjunction) that occurs for all D-state vectors elements χ and UDM(i) bits, observing conditions given in Table 3.4 [7].

If in the given D-state, D-action i is not realized, then ra$\chi(i) = 0$, otherwise ra$\chi(i) = 1$.

If it is known which D-actions are realized, then existence mark α_{is} can be recorded to all links, total capital investments and operation costs can be calculated, as well as system quality criterion in D-state $F(e(t))$ can be estimated.

3.3 Formation Methods of LTS D-Process

In the process of LTS development optimization, the correlation of chains of related D-states has been considered. Generally, technical systems are developing gradually by D-steps. If D-states $e(t-1)$ are developed to step $t-1$, then for step t, development extension $e(t)$ must be formed such that transition to it from any D-state $e(t-1)$ would be possible.

Let us assume expression

$$\chi(t-1) \rightarrow \chi(t), \tag{3.2}$$

which shows that the transition from D-state $e(t-1)$ to $e(t)$ is possible.

Expression (3.2) is equivalent to expression (3.3):

$$e(t) \in E[t, e(t-1) \rightarrow]. \tag{3.3}$$

The state of technical system graph G of all links (existing and alternative) depends on D-state vector χ_t. Probable transitions from any vector component i may be represented by the directed graph. Let us assume expression

$$M(t-1, i) \rightarrow M(t, i). \tag{3.4}$$

If transition is not possible, then let us use negation expression

$$M(t-1, i) \rightarrow \neg M(t, i). \tag{3.5}$$

In the example

$$1 \rightarrow 1,$$
$$0 \rightarrow 1,$$
$$0 \rightarrow 0,$$
$$1 \rightarrow \neg 0.$$

If D-action is realized (the fact has taken place), it can never be assumed as non-existent.

Transitions from $M(t-1, i)$ to $M(t, i)$ may also depend on other components, for example, on component j state. If $M(t, j)$, $M(t-1, i) = 0$ and $M(t, i) = 1$, then $M(t-1, i) \rightarrow \neg M(t, i)$, that is, D-actions i and j may not be realized concurrently in one D-state. Examples of such conditions were reviewed in Table 3.1. To take into consideration, this kind of situation, logical conditions of D-actions realization must be applied. In development optimization of LTSs, two types of logical conditions might be required.

Type 1 "Only one D-action from the set". For example, if the task is to be solved that addresses capacity of newly constructed links or selection of generation node capacity. For each condition, the number of D-actions is assigned in the set under consideration and index numbers of the D-actions, comprised in the set. In data array raχ, the number of realized D-actions m contained in the set is verified. If $m > 1$, then D-state $e(t)$ is not logically by admissible.

Type 2 "After". D-action j can be realized only in the case, if D-action i has been already realized. The second type of logical condition is applied if gradual construction of new elements is considered (construction takes place in several stages): (a) D-action j—first stage construction, (b) D-action i—second stage construction. Also verification basis for logical condition of type 2 is binary state vector χ of coding array raχ. D-state $e(t)$ is logically admissible, if ra$\chi(i) = 1$ and ra$\chi(j) = 1$, but is not logically admissible, if ra$\chi(i) = 1$ and ra$\chi(j) = 0$.

If for all D-state vector χ elements $i = 1, 2,...,n$ the condition $M(t - 1, i) \rightarrow M(t, i)$ is logically admissible and also all logical conditions are observed, then also

$$\chi(t - 1, i) \rightarrow \chi(t, j). \tag{3.6}$$

When optimizing LTS development [applying recursive equations and conditions (2.10), (2.21), (2.24) and (2.25)], it is required to verify several times, whether the transition from one D-state to another is possible.

When utilizing dynamic programming, the opportunity of the transition from the D-state in step $t - 1$ to another D-state in step t must be examined.

The verification performed according to condition (3.6) is not complicated. If D-states vectors $\chi(t - 1, 1)$ and $\chi(t, 2)$ are given, the verification is done in the following sequence:

1. Consider all D-state vector elements $i = 1, 2,...,n$;
2. For each element i compare D-state vector elements values $M(i, 1)$ and $M(i, 2)$ (see Table 3.5).

The condition that from the first D-state there may be the transition to the second $\chi(1) \rightarrow \chi(2)$ is fulfilled (not taking into consideration logical conditions) if all state vector elements $i = 1, 2,...,n$ meet the condition $M(i, 1) \rightarrow M(i, 2)$.

Condition (3.6) may be verified with this operation:

$$(\chi(t - 1, i) \vee \chi(t, j)) = \chi(t - 1, i), \tag{3.7}$$

where $\vee$—logical "or" (disjunction) that occurs for all D-state vector elements χ and UDM(i) bits, observing given conditions [1] in Table 3.4.

If condition (3.7) is fulfilled, then condition (3.6) is also observed—the transition is possible from D-state $\chi(t - 1, i)$ to D-state $\chi(t, j)$ is possible.

In Table 3.6, some examples are shown to illustrate how condition (3.7) is functioning.

	$M(i, 1)$	$M(i, 2)$	β
Table 3.5 Element i transition state vector possible ($\beta = +$) and impossible ($\beta = -$) values	0	0	+
	0	1	+
	1	1	+
	1	0	−

Table 3.6 Examples of condition (3.7) verifications

j	$\chi(1)_{10}$	$\chi(2)_{10}$	$\chi(1)\ M(i)$					$\chi(2)\ M(i)$					$\chi(1)\wedge\chi(2)\ M(i)$					Transition
			1	2	3	4	5	1	2	3	4	5	1	2	3	4	5	
1	0	27	0	0	0	0	0	1	1	0	1	1	0	0	0	0	0	There is
2	22	23	0	1	1	0	1	1	1	1	0	1	0	1	1	0	1	There is
3	22	30	0	1	1	0	1	0	1	1	1	1	0	1	1	0	1	There is
4	22	31	0	1	1	0	1	1	1	1	1	1	0	1	1	0	1	There is
5	22	14	0	1	1	0	1	0	1	1	1	0	0	1	1	0	0	There is no

j—example number; $\chi(1)10$—vector value in decimal system

Condition $\chi(1)\to\neg\chi(2)$ is also fulfilled if

$$\chi(1)<\chi(2). \tag{3.8}$$

Example 5 (see Table 3.6), $\chi(1)_{10} = 22$ and $\chi(2)_{10} = 14$; in accordance with condition (3.8), the transition is not possible, as value 22 is higher than 14.

3.4 Conclusion

The suggested LTS development formation algorithm is simple and fast.

Reference

1. Dale VA, Krishans ZP, Paegle OG (1969) Dynamic Programming in Calculation of Power Networks Development. Riga, Latvia (in Russian)
2. Dale VA, Krishans ZP, Paegle OG (1990) Dynamic Optimisation of Electric Power Networks Development. Zinatne, Riga, Latvia (in Russian)
3. Krishans ZP (1981) Architecture of dynamic models for power system networks optimal development. Изв. АН СССР, Энергетика и транспорт 5:32–41 (in Russian)
4. Krishans Z, Neimane V, Anderson G (1999) Dynamic Model for Planning of Reinforcement Investments in Distribution Networks. In: 13th Power Systems Computation Conference, Trondheim, Norway, Conference Proceedings, vol 2 pp 863–869
5. Krishans Z, Oleinikova I (2001) Software for Power Supply Utilities Management. Riga Technical University, Riga, Latvia (in Latvian)
6. Krishans Z, Oleinikova I, Mutule A (2007) Planning for Urban Medium Voltage Network. In:7th WSEAS/IASME International Conference on Electric Power System, High Voltages, Electric Machines (POWER'07), Venice, Italy (Conference Proceedings on CD 600–121)
7. Mendelson E (1997) Introduction to mathematical logic, 4th edn. Chapman & Hall, London, ISBN 978-0-412-80830-2

Chapter 4
LTS D-Process Dynamic Optimization on all D-States

Abstract This chapter is an introduction to OIS methods family analysis (Part II). Using the concepts from Chaps. 1–3 and LTS development process model, we perform development process plan optimization in all development steps saving all development states. The aims of this research work are: (1) to prove that the method reviewed is only valid for small technical system development optimization; (2) to create new LTS development optimization algorithms; (3) to explore these algorithms for finding new ways of method development. To achieve this goal we research system development process. The research has shown that in real development optimization tasks, it is necessary to consider only a small part of development states. This characteristic feature of optimization process gives opportunity to develop new dynamic optimization methods with a wider utilization range and a larger number of alternate development actions and development steps.

4.1 Recursive Equation Calculation Algorithm

This chapter discusses realization algorithm for LTS sustainable development optimization, using recursive Eq. 2.9. In the algorithm, we use the systems analyzed in Chap. 3 and their D-process modeling algorithms. Having observed that D-states $e(t)$ and $e(t-1)$ may be expressed with D-state binary vectors $\chi(t)$ and $\chi(t-1)$, let us rewrite recursive Eq. 2.9 as follows:

$$f[t, \chi(t)] = F[e(t)] + \max_{\chi(t-1) \to \chi(t)} f[t-1, \chi(t-1)] \qquad (4.1)$$

where $f[t,\chi(t)]$ is system objective function from the initial D-state $\chi(0,0)$ up to D-state $\chi(t)$ in D-step t; $f[t-1,\chi(t-1)]$ is system objective function from the initial D-state $\chi(0,0)$ up to D-state $\chi(t-1)$ in D-step $t-1$; and $F[e(t)]$ is D-state $e(t)$ quality criterion in D-step t.

Z. Krishans et al., *Dynamic Management of Sustainable Development*,
DOI: 10.1007/978-0-85729-062-5_4, © Springer-Verlag London Limited 2011

Table 4.1 Information structure in LTS development optimization, storing information on all D-states

t	i	Value	D-state				
	1	χ	0	1	2	...	$V-1$
1	2	$\pi[\chi(1)]$	a	a	a	...	a
	3	$f[1, \chi(1)]$	a	a	a	...	a
	4	optg $[\chi(1)]$	a	a	a	...	a
2	5	$\pi[\chi(2)]$	a	a	a	...	a
	6	$f[2, \chi(2)]$	a	a	a	...	a
	7	optg$[\chi(2)]$	a	a	a	...	a
...	...	...	...	...	...	...	...
T	8	$\pi[\chi(T)]$	a	a	a	...	a
	9	$f[T, \chi(T)]$	a	a	a	...	a
	10	optg$[\chi(T)]$	a	a	a	...	a

i Row number
[a] Table elements, where calculated values are recorded

Expression

$$\chi(t-1) \rightarrow \chi(t) \tag{4.2}$$

indicates that functional maximization occurs only in such D-states set, from which the transition to D-state $\chi(t)$ is possible.

LTS development optimization, based on recursive Eq. 4.1, is occurring gradually by filling out data arrays given in Table 4.1. Let us describe in what sequence optimization occurs. The first row of Table 4.1 contains records of all possible D-states that are expressed with D-state vector χ in the numerical form. χ is placed in the first line, from $\chi = 0$ and $\chi = 1$ in ascending order of χ numerical value up to $V-1$, where the number of D-states V is calculated by formula Eq. 2.26. Numerical χ value can be decoded applying diagonal matrix, and system D-state vector in data array raχ can be formed.

Optimization is started with first D-step $t = 1$, considering all D-state vectors χ, that are placed in the first row of Table 4.1 and for each χ as follows:

1. Calculate realized D-actions for D-state χ recorded in data array raχ, where ra$\chi(i) = 1$, if D-action i is realized, otherwise ra$\chi(i) = 0$.

   - Calculate summary investments and their costs per year.
   - Calculate operation costs that are caused by D-action realization.
   - Create system calculation graph in D-state χ, adding new links and disconnecting links to be removed; thus system parameters are changed and flow calculation can be performed.

2. Calculate output generation nodes.

   - Calculate transport flows of network links and overloading of links.

3. Verify the technical validity of χ: if system D-state is valid, then $\pi[\chi] = 1$, otherwise $\pi[\chi] = 0$.

4. Calculate system D-state quality criterion in the first D-step $F(e(1,\chi))$, taking into consideration investment in D-action realization and supplementary operation costs that are caused by D-action realization; if system D-state is invalid, then in calculating $F(e(1,\chi))$ penalty functions related to overloading and breach of other technical requirements must be taken into consideration.
5. Calculate functional $f[1,\chi]$; calculate the value of functional $f[1,\chi]$ with recursive Eq. 4.1; since for all D-states, in the first step there is only one initial D-state $\chi(0,0) = 0$, then

$$f[1,\chi(1)] = F(1,\chi(1)); \tag{4.3}$$

6. Record in the fourth row the initial D-state $\chi(t-1)$ of optimal development optg$[\chi(1)]$; note that in the first D-step for all D-states $\chi(t-1) = \chi(0,0)$.

After rows 2, 3 and 4 of the table are filled out for all D-states $\chi = 0,1,...,V-1$, then calculation of the first step is completed and its results (rows 2, 3 and 4) become the initial data of the second step.

Similarly, consider all D-states χ from 0 up to $V-1$ in the second D-step. For each χ:

1. Calculate realized D-actions data array raχ, total investments, system graph and its parameters.
2. Calculate manufactured production, flows and overloading of transport network links.
3. Calculate technical validity $\pi[\chi]$.
4. Calculate system D-state quality criterion in the first D-step $F(e(2,\chi))$.

```
begin;
maxft:=maxf(t-1); opt g [χ(t)]=0;
for i:=0 to V-1 do begin;
for j:=0 to V-1 do begin;
if f(t-1,j)→χ(t,i) then
if f(t-1,j)>maxft then begin;
maxft:=f(t-1,j);
opt g [χ(t)]:=j;
end;
end;
f(t,χ(i))=F(e(t))+maxft;
end;
```

Fig. 4.1 Recursive Eq. 4.1 calculation algorithm

5. Calculate functional $f[2,\chi]$; calculation algorithm for functional $f[2,\chi]$ in the sixth row differs from the first step algorithm—this already utilizes to full extent recursive Eq. 4.1.

Functional $f[t - 1,\chi(t - 1)]$ maximization algorithm (see Fig. 4.1) is simple; it is based on all D-states $\chi(t - 1)$ considered consecutively (Fig. 4.1).

The second step calculation is completed if rows 5, 6 and 7 are completed for all D-states.

The calculation of the third, fourth and all other steps including D-step T, is performed in the same way as in the second D-step.

When all steps are considered, optimization result corresponds to functional $f[T,\chi(T)]$ maximal value in the ninth row. Optimal D-process can be calculated in reverse mode—undo step t: $= T, t - 1,..., 2, 1$, utilizing data array optg($e(t)$) with t values $T, t - 1,...,2, 1$.

4.2 Algorithm Programming Aspects

The algorithm discussed in Sect. 4.1 provides opportunity to elaborate software for LTS development optimization. The creation of D-states with χ index-consecutive numbering ensures opportunity not to store D-states in computer memory (see Table 4.1, first row). It is sufficient to provide some space for two D-steps $t - 1$ and t. The stored data arrays required for dynamic optimization are as follows.

$$
\begin{aligned}
&\pi[\chi(t - 1)], \\
&\pi[\chi(t)], \\
f[t - 1&, \chi(t - 1)], \\
&f[t, \chi(t)].
\end{aligned}
\tag{4.4}
$$

The length of each data array stored is V.

In transition from recurrent step t to the next step $t + 1$, step t data array content is forwarded to step $t - 1$ data array.

When data arrays Eq. 4.4 are only used, system development can be optimized; however, in this way we obtained data only on the system latest D-state opt$\chi(T)$. The latter determines which D-actions must be realized but we are not aware when these must be realized. We are also not aware that the most optimal D-process is within the entire optimization period.

In order to print out complete D-plans description either on optimal D-plan or competitive D-plans including realized D-actions and realization time, information on optimal initial state in the previous step optg($e(t)$) must be stored for all steps and D-states in each step. For storing one initial state, two bites of computer memory are required. The scope of initial D-states to be stored is calculated by the following formula:

Table 4.2 Computer random access memory required for optimal and other D-processes storage

No.	n	T	Random access memory
1	10	10	10 kb
2	16	10	60 kb
3	20	20	400 mb

Fig. 4.2 Software architecture for LTS development optimization

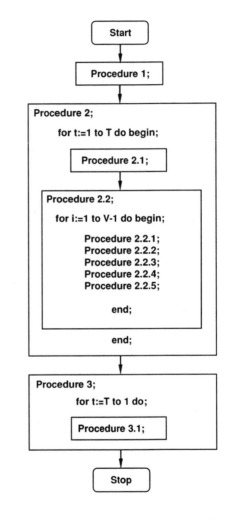

$$V = 2^n \times T, \tag{4.5}$$

where n is the number of D-actions and T is the number of D-steps. Capacity figures of memory required for storing D-plan are presented in Table 4.2.

Modern computers have enough memory to store the information required. The structure and architecture of LTS development optimization software is depicted in Fig. 4.2.

Table 4.3 Optimization results

D-plans	1	2	...	10
Quality criterion	[a]	[a]	...	[a]
Realized D-actions				
D-action 1[b]	[b]	[b]	...	[b]
D-action 2	[b]	[b]	...	[b]
...	...	...	...	...
D-action n	[b]	[b]	...	[b]

[a] D-plan quality criterion value
[b] D-action realization D-step in specific D-plan; if D-action is not realized, in the respective box there is "–"

The software consists of 11 major procedures (software blocks).

Procedure 1	Date entry from database.
Procedure 2	Optimization process cycle by D-steps.
Procedure 2.1	D-step t data input, for example, consumptions.
Procedure 2.2	Cycle by D-steps, calculation of system indicators for all D-states (see Table 4.1).
Procedure 2.2.1	$\chi := i$, decoding of χ in data array ra$\chi(n)$, utilizing unit diagonal matrix UDM(n). Calculation of investments. Creation of system calculation graph for D-state χ.
Procedure 2.2.2	Calculation of generation nodes capacity, transport network links flows and overloads.
Procedure 2.2.3	Determination of D-state technical validity $\pi[\chi]$.
Procedure 2.2.4	Calculation of technical system D-state quality criterion in D-step t and in D-state χ-$F(\chi(t))$.
Procedure 2.2.5	Calculation of functional for D-state $\chi(t)$: if $t = 1$, then $f[1,\chi(i)] = F[1,\chi(i)]$; if $t > 1$, then $f[t,\chi(i)]$ is calculated using procedure illustrated in Fig. 4.1.
Procedure 3	In undo D-step for 10 best D-plans, D-processes $\chi(T)$, $\chi(t-1)$, ..., $\chi(t)$, $\chi(t-1)$,..., $\chi(1)$ are formed.
Procedure 3.1	$\chi(t)$ is determined.

Optimization results are extracted in the form of a table for 10 D-plans (see Table 4.3).

4.3 System D-Process

In order to estimate the advantages of dynamic programming, let us review development optimization process using routine electric network development optimization as an example (see Fig. 4.3) [1, 2].

Fig. 4.3 System calculation graph

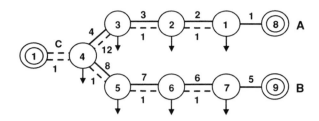

Table 4.4 D-states example

χ		D-action
Numeral	Vector	
0	000	System initial D-state
1	001	Graph loop 1–4 reconstruction
2	010	Graph loop 4–4 reconstruction
3	011	D-actions 1 and 2 joint realization
4	100	New generation node 10 construction
5	101	D-actions 1 and 4 joint realization
6	110	D-actions 2 and 4 joint realization
7	111	D-actions 1, 2 and 4 joint realization

Table 4.5 Optimization example results

D-plan	$F(T = 5)$, m.u.	$\chi(5)$	optg				
			$\chi(1)$	$\chi(2)$	$\chi(3)$	$\chi(4)$	$\chi(5)$
1	−1797	111	010	010	011	111	–
2	−181.7	110	010	010	110	110	110
3	−205.7	101	001	101	101	101	101
4	−208.8	100	100	100	100	100	100

In the above example, eight D-states are considered (see Table 4.4).

Here, optimization is reviewed for a 20 year estimation period that is structured into five D-steps. It is assumed that in the example considered, in all D-plans the quantity and price of the product to be sold is equal. Under such conditions, total reduced operation costs can be used as an optimization criterion.

$$F(T, e(T)) = \sum_{t=1}^{T} F(e(t)). \tag{4.6}$$

Optimization results are presented in Table 4.5.

In the example, the system initial D-state already in the first step is technically invalid. Therefore, in the first D-step the system must be reconstructed—proper D-action must be selected.

Optimization results analysis proved that in the example provided, the first D-action can only be selected if system development further process is estimated

Table 4.6 D-plans by t D-steps from $\chi(0)$ up to $\chi(t)$

t	$\chi(t)$							
	0	1	2	3	4	5	6	7
1	2	3	4	5	6	7	8	9
1	$=$	$\frac{0}{1}$	$\frac{0}{1}$	$\frac{0}{1}$	$\frac{0}{1}$	$\frac{0}{1}$	$\frac{0}{1}$	$\frac{0}{1}$
2	$=$	$=$	$\frac{2}{1}$	$\frac{1,2,3}{1}$	$\frac{4}{1}$	$\frac{1,4,5}{3}$	$\frac{2,4,6}{3}$	$\frac{1,2,3,4,5,6,7}{7}$
3	$=$	$=$	$=$	$\frac{2,3}{4}$	$\frac{4}{1}$	$\frac{4,5}{4}$	$\frac{2,4,6}{5}$	$\frac{2,3,4,5,6,7}{18}$
4	$=$	$=$	$=$	$=$	$\frac{4}{1}$	$\frac{4,5}{5}$	$\frac{4,6}{6}$	$\frac{3,4,5,6,7}{33}$
5	$=$	$=$	$=$	$=$	$\frac{4}{1}$	$\frac{4,5}{6}$	$\frac{4,6}{7}$	$\frac{4,5,6,7}{44}$

Technically non-valid D-state

and considered. For instance, taking into consideration only the first D-step, the most advantageous D-action is construction of new power supply source, not reinforcing network links (D-plan 4). Still, in the fourth D-plan expenditures up to the fourth D-step are higher than in the second D-plan with D-actions 3 and 2. Therefore, expenditures of one D-step insufficiently justify the decision made.

Optimization example evidently demonstrates that estimation period must be sufficiently long as excess expenditures at the beginning may be paid off in the future.

The significant fact is that expenditures in D-state in the dated step depend only on realized D-actions set and not on when the D-actions are realized, i.e., $F(e,t)$ does not depend on system development up to the step t. If in D-process D-actions costs are changed, then one D-action must be modeled with several D-actions, where each D-action is planned with its own realization period allowed.

Let us review how many D-plans are possible in the example. In the example, there are three, not interrelated, independent D-actions; only one type of each D-action can be realized. Each D-action can be realized in D-step 1, 2, 3, 4 and 5

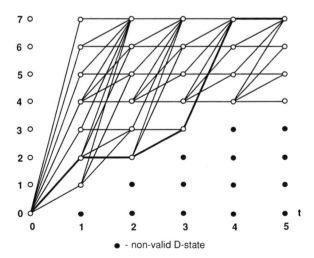

Fig. 4.4 System example development graph and optimal D-plan, if $T = 5$

• - non-valid D-state

or may be not realized at all. Thus, theoretically the number of possible D-plans is 216. In order to calculate all prospective D-plans in sequence, 1,080 D-states must be estimated and calculated. Utilizing dynamic programming method only 10 D-states must be calculated. Therefore, in this simple example also dynamic programming is more effective than the revision of all D-plans. By increasing the number of D-actions, it is not possible to review all D-plans.

Optimization process, based on dynamic programming is illustrated in Table 4.6 and Fig. 4.4.

In Table 4.6, the following values are given: in numerator part—D-states $\chi(t-1)$ (numerical values), from which may be the transition to D-state $\chi(t)$. Optimal initial D-state is indicated with bold digit. In denominator part, the number of possible D-plans is given.

Table 4.6 also provides information on optimal development from $\chi(0)$ up to $\chi(t)$. For instance, in order to determine optimal development by three D-steps up to $\chi(t=3)=6$, in undo step we can determine $\chi(t=2)=2$ and $\chi(t=1)=2$.

Let us analyze the example of optimization process in detail [3, 4]. In the first D-step $t=1$, there are seven technically valid D-states. The existing D-state is technically invalid. In any D-state, the first D-step can be reached only from the existing D-state in D-step 0. In the first D-step, only first step D-state technical and economic criteria as well as D-states quality criteria are calculated, but optimization does not occur.

In the second D-step $t=2$, all probable second step D-states are analyzed. In the second step, there are six technically valid D-states. In the second D-state, there is only one possible D-plan from the second D-step. This is the optimal D-plan.

In the third D-state, the second D-step can be reached from D-states 1, 2 and 3 in the first D-step. Optimal development by total reduced system quality criterion by two D-steps is the transition from the second D-state. The two remaining D-plans further on are not considered (these are abandoned) as already in the second D-step it is known that these are worse than others; so these cannot be optimal D-plans. Such abandonment process of D-plan is typical for dynamic programming.

Similarly, on reviewing D-states 4, 5, 6 and 7 in the second D-step, all those D-plans are abandoned which in the first two D-steps are these D-states sequences: 4,5; 5,5; 4,6; 6,6; 1,7; 3,7; 4,7; 5,7; 6,7 and 7,7. Thus, in the next D-steps, only those D-plans are considered, which in the first two steps contain one of the following D-states sequences: 1,5; 2,2; 2,3; 2,6; 2,7 and 4,4.

Having considered development graph in our example (see Fig. 4.4), we come to the conclusion that out of 18 graph loops, accordingly $\chi=0$, $t=0$, in the second D-step remain only 6.

In the third D-step, $t=3$ five possible D-states development routes are analyzed. The third D-state can be reached by four routes: 1,3,3; 2,2,3; 2,3,3 and 3,3,5. But D-plans which in the first two D-steps contain D-states sequences 1,3 and 3,3 are already abandoned in the second D-step. Consequently, in the third D-step, for D-state 3 only two D-plans remain: 2,2,3 and 2,3,3. Optimal D-plan by three

D-steps is 2,2,3. This D-plan is stored, but D-plan 2,3,3 is abandoned (is not considered in further optimization).

Optimal D-plan calculation for other D-states of the third D-step occurs in similar way. Out of 32 graph loops that begin at $\chi = 0$, $t = 0$ and end at $\chi = 0$, 1,...,7, $t = 3$, only 5 are preserved for further optimization process; thus the number of D-steps increases, the efficiency of dynamic programming is growing considerably.

In optimizing further steps, similar process occurs in the result of which, in the last fifth optimization step, out of 58 possible D-plans only 4 remain amid optimal D-plan too.

References

1. Dale VA, Krishans ZP, Paegle OG (1970) Transmission network development optimization by dynamic programming. Изв. АН СССР, Энергетика и транспорт 4:91–96 (in Russian)
2. Dale VA, Krishans ZP, Paegle OG (1979) Dynamic methods for analysis of power system networks development. Zinatne, Riga, Latvia (in Russian)
3. Krishans Z (1998) Methods of modelling and optimization for power supply utilities management. Riga Technical University, Riga, Latvia (in Latvian)
4. Neimane V, Anderson G, Krishans Z (1998) Multi-criteria analysis application in distribution network planning. In: 33rd Universities power engineering conference UPEC'98, (conference proceedings) Napier University, Edinburgh, UK vol. 2 pp 791–794

Chapter 5
Optimal Initial States Method

Abstract This chapter describes the analysis of optimal initial state methods family basics, optimality principles and recursive equation. In this chapter, we define optimal initial state set characteristics and formation principles. Taking into consideration OIS methods' general characteristics, we offer OIS method's algorithm, which can be adjusted according to system properties. Also, there is reviewed, the data structure of optimal initial states stored. There are analyzed optimization process characteristics, utilizing OIS methods. There are different researched formation methods of development state steps. For determination of optimization program, optimal initial states set variable, various development states formation algorithms can be used. For example, if development states vectors are binary digits, but development states are formed adding one unit (action), then such algorithm is of considerable drawback—technical system graph formation process is occasional and such algorithm can only be utilized if development actions number is not large. The most effective formation methods of optimal initial states are presented in detail in Chap. 6.

5.1 Dynamic Programming Application for LTS Sustainable Development Management

Dynamic programming is widely used to solve tasks of diversified power systems development planning [1–24]. When comparing with other methods, the advantage of this method is its ability to tackle different practical tasks, in particular, if a task is subject to a lot of constraints of various nature and if variables are discrete values, but objective function is non-differential integer nonlinear function. The authors of the book are of the opinion that there are similar conditions for other LTSs, and suggest sharing their experience and applying the methods for systems sustainable development management.

Dynamic programming is intended to solve tasks on multi-step process management. The basic idea of dynamic programming is worked out by Bellman

[25, 26]. He has formulated general characteristics of multi-step solution process. Bellman's principle of optimality states: the property of an optimal process is that whatever the initial state and initially assumed values are; further optimal decisions are made with regard to the state resulted from the first decision made. The solution of various concrete tasks with dynamic programming is taking place according to the unified scheme. The task is considered as a T steps decision-making process during which a decision is made for the current step t. The significant structural characteristic of this task is its form independence of the steps number, so decision-making algorithm execution does not depend on the number of steps. In such a case, the tasks with different steps number are solved in sequence. In this process, the decision is obtained on step's t task, supplementing step's $t - 1$ solution. The essential characteristic of dynamic programming technique is recursive equation utilization, as well as objective function additivity (objective function must be expressed as a sum); state criterion must not depend on how the existing state is achieved. This condition ensures the opportunity to utilize dynamic programming optimality principle.

Optimality principle of LTSs sustainable development management in case of optimization is formulated as follows: a technical system in optimal development time period $t + 1,...,T$ from D-state $e(t)$ up to D-state $e(t)$ does not depend on D-process from $e(0)$ to $e(t)$ and such optimality principle formulation allows composing a recursive equation, with whose help the maximal value of objective function $f(t,e(t))$ can be calculated for D-process g from D-state $e(0)$ to D-state $e(t)$ in time period t:

$$f[t, e(0)] = \max_{\{g(t,e(0))\}} [F(e(1)) + \cdots + F(e(T))], \tag{5.1}$$

where $\{g(t, e(t))\}$, admissible D-processes set in time period t up to D-state $e(t)$; $F(e(t))$, system state $e(t)$ in D-step t quality criterion.

Objective function (5.1) maximization is illustrated in Fig. 5.1; in this figure four possible LTS D-processes are presented.

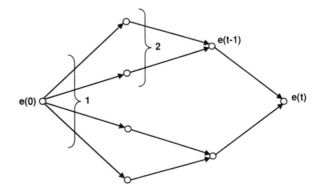

Fig. 5.1 Illustration of recursive equation formation to maximize objective function *1* D-process up to $e(t)$ set, *2* D-process up to $e(t - 1)$ set

Let us mark with $\{e(t-1) \subset e(t)\}$ D-states $e(t-1)$ set, from which transition is possible to D-state $e(t)$, but with $\{g(t-1, e(t-1))\}$—D-process from D-state $e(0)$ to D-state $e(t-1)$ in time period $t-1$.

Objective function maximization may be performed in the following sequence:

$$f[t, e(t)] = \max_{e(t-1) \subset e(t)} F(e(t)) + \max_{\{g(t-1, e(t-1))\}} [F(e(0)) + F(e(1)) + \cdots + F(e(t-1))].$$

$$(5.2)$$

In our model, $F(e(t))$ is independent of $e(t-1)$. Besides,

$$\begin{aligned} &f(t-1, e(t-1)) \\ &= \max_{\{g(t-1, e(t-1))\}} [F(e(0)) + F(e(1)) + \cdots + F(e(t-1))]. \end{aligned} \qquad (5.3)$$

Thus,

$$f[t, e(t)] = F(e(t)) + \max_{e(t-1) \subset e(t)} f[(t-1, e(t-1))]. \qquad (5.4)$$

Expression (5.4) is dynamic programming recursive equation that enables one to create multi-step optimization process for time period T from $e(0)$ to any $e(T)$. Optimization procedure is performed in the inductive way starting from the first D-step and ending with the last one: if $t = 1$, then $f(1, e(1)) = F(e(t))$.

Algorithm for objective function maximization in D-step t. All probable D-states are to be reviewed in sequence $e(t) \in \{e(t)\}$. For all $e(t)$, the following operations are to be fulfilled:

1. Calculate technical, economic, production supply reliability and ecological criteria as well as D-state $e(t)$ in step t quality criterion $F(e(t))$.
2. Determine D-states $e(-1)$ set, from which the transition is possible to D-state $e(t)$ (liabilities verification algorithms are considered in Chap. 3).
3. With the help of recursive equation (5.4) determine D-state $e(t)$ functional $f(t, e(t))$ maximal value.

To apply the algorithm, functionals $f(t-1, e(t-1))$ and $f(t, e(t))$ must be stored for all D-states (algorithm utilization technique is reviewed in detail in Chap. 4).

Recursive equation (5.4) can only be used in tasks with a small number of D-actions (up to 10): as the number of D-actions increases, the scope of possible D-states grows exponentially. Respectively, calculation time and computer memory capacity required are growing too. Thus, optimization methods are required for LTS sustainable development management with the increased alternative D-actions number $n = 30$ and such methods are reviewed in this chapter and Chap. 6.

5.2 Recursive Equation for Optimal Initial State Method

To solve the problem of computational resources (time and memory) in multi-step optimization tasks, attempts have been made to use specialized procedures, for example, random search method (Monte Carlo) and truncated dynamic programming [8, 27]. The methods elaborated by the Laboratory for Power System Mathematical Modeling (PSMM) have supplemented optimality principle by observing logical regularities consequences of LTS D-process, which is reviewed in detail in Chap. 6. Optimization methods can also be found which are more effective than recursive equation (5.4) direct realization (see Chap. 4).

In step t for all D-states $e(t)$, there is only one optimal development $optg(e(t))$ from D-state $e(0)$ up to D-state $e(t)$, as well as initial D-state, which will be marked with $\omega(t - 1)$. Let us mark with $\Omega(t - 1)$ initial D-state $\omega(t - 1)$ set, from which an optimal transition is possible to any D-state $e(t)$. Let us mark $\Omega(t - 1)$ as optimal initial states (OIS) set. Development optimization methods that utilize OIS set will be called OIS methods.

It should be noted that specific D-actions efficiency depends on system D-state and time moment. Owing to advantageous structural specific characteristics of OIS methods, there is no need to store all D-states $e(t) \in \{e(t)\}$. Still, OIS methods can only be applied if it is possible to determine OIS set $\Omega(t)$ for unknown functional $f(t + 1, e(t + 1))$ values in D-step $t + 1$. This kind of determination method also exists.

Functionals $f(t,e(t))$ must be stored only for OIS set sequence: $\Omega(1)$, $\Omega(2)$, ..., $\Omega(t)$, ..., $\Omega(t - 1)$ that provides significant economy of computer memory, because in real technical system development optimization, tasks set $\Omega(t)$ is considerably less than set $\{e(t)\}$. This is due to local reasons which cause the necessity of system development. As a rule, D-actions are effective only in some D-steps as system development causes are structured in time and space.

According to the conditions of technical system D-process, g, $\{e(t - 1)\} \subset \{e(t)\}$. Transitions from $\{e(t-)\}$ D-states to $\{e(t)\}$ D-states describe system D-process in time interval $t - 1$, t; let us mark them with $e(t - 1) \subset e(t)$. For D-states set $\{e(t)\}$ let us extract subset $\Omega(t) \subset \{e(t)\}$. The properties of D-states $\omega(t)$ of set $\Omega(t)$ are defined as follows:

1. In set $\Omega(t)$ for any D-state $e \in \{e(t)\}$ possible transition can be determined from D-state $\omega(t)$ to D-state e.
2. If $\omega(t) \subset e$, then $e \subset \neg\Omega(t)$ if condition $f(t,\omega(t)) \geq f(t, e)$ is observed, where $f(t, e)$ is the maximal value of objective function (system development from D-state $e(0)$ to D-state e within time period t).

Functional $f(t, e)$ value can be determined as a result of t D-steps optimization process, utilizing recursive equation (5.4). Let us mark with $\varphi(t, e)$ the maximal objective function value, which we obtain in optimization process of t D-steps, by using recursive equation (5.5):

$$\phi(t, e(t)) = F(e(t)) + \max_{\{\omega(t-1) \subset e(t)\}} f(t-1, \omega(t-1)). \qquad (5.5)$$

Recursive equations (5.5) and (5.4) differ only in that maximization according to (5.5) occurs within limits of OIS set, while in (5.4) it occurs within D-states $\{e(t-1) \subset e(t)\}$ set limits. It can be assured that objective function maximal value is not varying, therefore $\varphi(t, e) = f(t, e)$. To prove it, let us assume the opposite:

$$\phi(t, e) < f(t, e). \qquad (5.6)$$

In such a case, there must be D-state $e(t-1) \subset e$ that belongs to the set $\{e(t-1)\}$, but does not belong to set $\Omega(t-1)$, which means that

$$f(t-1, e(t-1)) > \max_{\{\omega(t-1) \subset e\}} f(t-1, \omega(t-1)). \qquad (5.7)$$

However, according to the first property of optimal initial state set $\Omega(t-1)$ there must be at least one initial state $\omega(t-1) \subset e(t-1)$. Then, in accordance with the second property of optimal initial state we have:

$$f(t-1, \omega(t-1)) \geq f(t-1, e(t-1)). \qquad (5.8)$$

Condition (5.8) is opposite to condition (5.7), that is, why $\varphi(t, e) = f(t, e)$.

This property also proves that initial states relevancy to the first and second property of optimal initial states is sufficient to determine OIS.

A general method for OIS determination in D-step t is as follows: D-state $\omega(t-1)$ is determined with the help of the following condition:

$$f(t, \omega(t, 1)) = \max_{\{e(t)\}} f(t, e(t)). \qquad (5.9)$$

It is obvious that $\omega(t, 1) \subset \Omega(t)$, as there is no such $e(t)$, for which

$$f(t, e(t)) \geq f(t, \omega(t, 1)). \qquad (5.10)$$

From (5.10) and second property of OIS it follows that all D-states $e(t)$, observing condition $\omega(t, 1) \subset e(t)$, cannot be OIS.

The next optimal initial state can be determined using condition (5.11):

$$f(t, \omega(t, 2)) = \max_{\{e(t,1)\}} f(t, e(t, 1)), \qquad (5.11)$$

where $\{e(t, 1)\} = \{e(t)\} - \{\omega(t, 1) \subset e(t)\}$; from optimization area there are excluded all those $e(t)$, to which one can pass from $\omega(t, 1)$.

It is obvious that $\omega(t, 2)$ is the most optimal initial state, although $f(t, \omega(t, 2) \leq f(t, \omega(t, 1)$; still $\omega(t, 1) \subset \neg\omega(t, 2)$. Besides, in set $\{e(t, 1)\}$ there is no D-state that observes condition $f(t, e(t) \geq f(t, \omega(t, 2)$.

If optimal initial states sequence $\omega(t, 1)$, $\omega(t, 2), \omega(t, 3), \ldots, \omega(t, i-1)$ is determined, then next optimal initial state $\omega(t-1)$ can be obtained using condition

$$f(t, \omega(t, i)) = \max_{\{e(t,i)\}} f(t, e(t, i)), \qquad (5.12)$$

where $\{e(t, i)\}$—maximization area.

For all D-states $e(t, i)$, these $i - 1$ conditions must be observed:

$$\omega(t, 1) \subset \neg e(t, i),$$
$$\omega(t, 2) \subset \neg e(t, i),$$
$$\cdots \qquad\qquad (5.13)$$
$$\omega(t, i - 1) \subset \neg e(t, i).$$

If determination of OIS is continued $\{e(t, i)\} = 0$, then all $\omega(t, i)$ in D-step t are determined.

The concept of the *most optimal initial state* and its determination general method are illustrated in Fig. 5.2. LTS D-states set $\{e(t)\} = \{1, 2, 3, 4, 5, 6, 7, 8\}$ is shown as a guided graph. D-states $e(t, i)$ and $e(t, j)$ that are connected with graph edges directed from $e(t, i)$ to $e(t, j)$, meet condition $e(t, i) \subset e(t, j)$.

The first optimal initial state $\omega(t, 1)$ is determined, employing condition

$$f(t, \omega(t, 1)) = \max_{\{e(t)\}} f(t, e(t)). \qquad (5.14)$$

Let us assume that $\omega(t, 1) = 3$. To obtain the second optimal initial state $\omega(t, 2)$, this condition is used:

$$f(t, \omega(t, 2)) = \max_{\{e(t,1)\}} f(t, e(t, 1)), \qquad (5.15)$$

where maximization area is $\{e(t, 1)\} = \{1, 2, 4\}$, which corresponds to condition $\omega(t, 1) \subset \neg e(t, 1)$. Let us assume that $\omega(t, 2) = 2$. Thus $\omega(t, 3)$ is determined utilizing condition

Fig. 5.2 Illustration of OIS determination general method

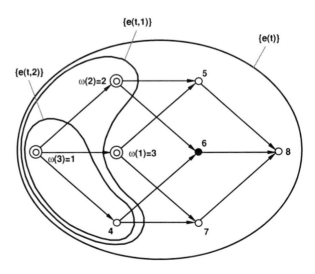

$$f(t, \omega(t,3)) = \max_{\{e(t,2)\}} f(t, e(t, 2)), \tag{5.16}$$

where maximization area is $\{e(t, 2)\} = \{1, 4\}$.

Let us assume that $\omega(t, 3) = 1$, where **1** is the initial state. Thus $\omega(t, 3)$ is the latest optimal initial state.

Recursive equation (5.5) allows one to optimize LTS development from $e(0)$ up to $e(T)$ within time period T, storing only $\Omega(t - 1)$ and $\Omega(t)$, $f(t - 1, \omega(t))$ and $f(t, \omega(t))$ in optimization process.

Optimal initial states recursive equation

$$f(t, e) = F(e(t)) + \max_{\{\omega(t-1) \subset e\}} f(t - 1, \omega(t - 1)) \tag{5.17}$$

allows to work out algorithms that are appropriate for real practical tasks solution on LTS development optimization.

Fig. 5.3 Flowchart for LTS development optimization, applying OIS method *mt* max number of realized D-action in D-step t

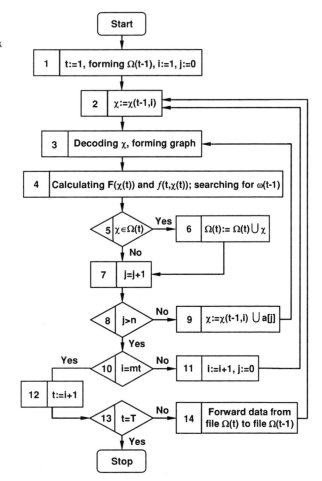

5.3 Algorithmization Principles of OIS Method

OIS method can be realized in various optimization programs and such software algorithm structure is represented schematically in Fig. 5.3. Block main function is to calculate, in step t, for D-states step quality criterion $F(e(t))$ for functional $F(t, e(t))$ development from $e(0)$ to $e(T)$.

LTS development optimization process is cyclic in D-steps $t = 1,\dots,T$, where the result of calculation in step t, moving to the next step $t + 1$, becomes its input data. D-step t initial data and results structure is given in Table 5.1. It illustrates OIS set of D-process in steps $t - 1$ and t.

Table 5.1 OIS method data structure

	i	Optimal initial state											Functional	$i(t - 1)$
Input data														
$t - 1$	1	1	1	0	0	0	0	0	0	0	0	0	115.2	
	2	1	0	0	0	1	0	0	0	0	0	0	115.0	
	3	1	0	0	0	0	0	0	0	0	0	0	114.9	
	4	0	0	0	1	1	0	0	1	0	0	0	113.7	
	5	0	0	0	1	1	0	0	0	0	0	0	113.7	
	6	0	0	0	0	1	0	0	0	0	0	0	113.5	
	7	0	1	1	0	0	0	0	0	0	0	0	112.8	
	8	0	0	1	0	0	0	0	1	1	0	0	112.1	
	9	0	0	1	0	0	0	1	1	0	0	0	112.1	
	10	0	0	1	0	0	0	0	1	0	0	0	112.0	
	11	0	0	1	0	0	0	0	0	1	0	0	111.6	
	12	0	0	1	0	0	0	0	0	0	0	0	111.6	
	13	0	0	0	0	0	1	1	0	0	0	0	111.5	
	14	0	0	0	0	0	1	0	0	0	0	0	109.3	
	15	0	0	0	0	0	0	1	0	0	0	0	107.3	
	16	0	0	0	0	0	0	0	0	0	0	0	94.8	
Results														
t	1	1	0	0	0	0	0	0	0	0	0	0	236.6	3
	2	0	0	0	0	1	1	0	0	0	0	0	234.1	6
	3	0	0	0	0	1	0	0	0	0	0	0	232.9	6
	4	0	0	1	0	0	0	0	1	0	0	0	232.5	10
	5	0	0	1	0	0	0	0	0	1	0	0	232.1	11
	6	0	0	0	0	0	1	1	0	0	0	0	231.7	13
	7	0	0	1	0	0	0	0	0	0	1	0	231.6	12
	8	0	0	1	0	0	0	1	0	0	0	0	231.5	12
	9	0	0	1	0	0	1	0	0	0	0	0	231.3	12
	10	0	1	1	0	0	0	0	0	0	0	0	230.8	7
	11	0	0	1	0	0	0	0	0	0	0	0	230.3	12
	12	0	0	0	0	0	1	0	0	0	0	0	229.8	14
	13	0	0	0	0	0	0	1	0	0	0	0	227.0	15
	14	0	0	0	0	0	0	0	0	0	0	0	211.9	16

It is recommended to employ an algorithm that arranges OIS in descending order beginning from functional maximal value and such algorithm is used in processing the data presented in Table 5.1. Then, index of consecutive numbering $i = 1$ indicates that functional $f(t, \omega(t, 1))$ is maximal, $f(t, \omega(t, 2))$ is the second with regard to functional maximal value and so on.

OIS disposition in descending sequence of functional $f(t, \omega(t, i))$ values makes execution of various frequently repeated operations fast and easy, for instance, when searching for an optimal initial state $\omega(t - 1) \subset e$ (see Fig. 5.3, block 4), this simplified condition can be used:

$$\omega(t - 1, i) \subset e \, i = 1, 2, \ldots, i_{\min}, \tag{5.18}$$

where $i_{\min}$—the least index number of consecutive numbering for optimal initial state $\omega(t - 1)$, from which there is transition to D-state e.

For example, the transition to D-state $\omega(t) = 0010010000$ is possible from D-states $\omega(t - 1)$, which are recorded in rows with index number $i(t - 1) = 12$ and $i(t - 1) = 14$. Optimal transition corresponds to the highest functional $f(t - 1, \omega(t - 1))$ value, i.e., $i_{\min} = 12$.

LTS D-steps optimization results are presented in Table 5.1. In such data arrays, OIS $\omega(t)$ and functionals values that correspond to optimal D-process up to $\omega(t)$ are recorded. Besides, in such data, arrays other demanded values can be recorded as well.

As shown in Chap. 4, D-states may be expressed with binary vectors χ, whose components are binary digits 0 or 1. That structural form was used to represent OIS $\omega(t)$ in Table 5.1. In specific software, the data format must be utilized, which makes it possible to consider optimal initial state $\omega(t)$ too as a binary vector χ. That, in turn, ensures compact D-state recording in computer memory and fast data processing that is particularly significant due to a capacious amount of information.

In D-step $t = 0$ only one row is recorded: $i(0) = 1$, $\omega(0) = 0$, $f(0, \omega(0)) = 0$. In step $t = T$ in the first row, there are data on optimal D-process by T D-steps from $e(0)$ to $\omega(T, i = 1)$. Objective function $f(T, \omega(T, i = 1))$ value is global maximum and is at the same time optimization task solution.

A characteristic feature of dynamic programming is obtaining not only optimal solution $i = 1$, but also other competitive solutions up to all OIS $\omega(t, 1)$ in step T.

If optimization result is obtained in the form of data array rows for all D-steps $t = 0, 1, 2, \ldots, T$, then in undo step D-states optimal sequence $e(0), \omega(1), \omega(2), \ldots, \omega(t, i)$ can be determined, which unambiguously corresponds to optimal D-actions realization sequence and realization moment.

OIS number W_M is usually changing by D-steps (see Fig. 5.4). The illustration presents a 20-step optimization task. The OIS number of optimization process is changed from 1 to 56. When comparing with total D-states number 800, the OIS number is not considerable. Still, utilizing OIS method in practice, it has been found that OIS data array repletion is probable that would consequently cause a loss of worse OIS. According to Fig. 5.4, D-process D-states indexes are not high—in the example reviewed the maximal index does not exceed 8.

Fig. 5.4 Optimization process characteristic curves, utilizing OIS method. *1* OIS number W_M, *2* optimal D-process index

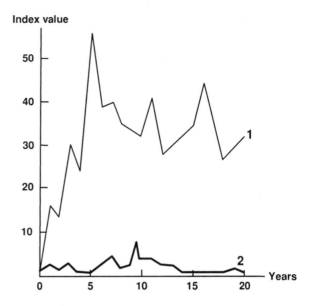

In the example of another research, the maximal index was 10. In practical tasks, not losing global optimum, it is sufficient with data array maximal lines number up to 50. The issue on limited table utilization will be reviewed in detail in Chap. 7.

OIS set in D-step t can be determined by various methods, for instance, by iteration method. From D-states set $\{e\}$ in we select D-states e_v, $v = 1, 2,...,v_M$ a certain sequence. Let us mark with $\{e\}_v$ D-states set that comprises D-states e_1, $e_2,...,e_v$, but with $\Omega(t, v)$—OIS set that is determined from $\{e\}_v$.

If $v = 0$, then set $\Omega(t, v)$ is empty. The first iteration ($v = 1$), $\{e\}_{v=1}$ comprises a single D-state $e_{v=1}$. Thus, also OIS set $\Omega(t, v)$ comprises only one optimal initial state

$$\omega = e_{v=1}. \tag{5.19}$$

In some iteration v, where $v > 1$, let us select D-state e_v, for which we calculate functional $f(t, e(v))$. D-state e_v belongs to the set $\Omega(t, v)$, if in the set there is no such optimal initial state, which concurrently corresponds to the following conditions:

$$f(t, \omega \subset e(t, v)) > f(t, e(t, v)). \tag{5.20}$$

If $e(t, v) \in \Omega(t, v)$, then $\Omega(t, v)$, besides $e(t, v)$, comprises also state $\omega \in \Omega(t, v - 1)$, which corresponds to the condition

$$e(t, v) \subset \neg\omega(t) \tag{5.21}$$

or

$$e(t, v) \subset \omega, \ f(t, e(t, v)) < f(t, \omega(t)). \tag{5.22}$$

If $e(t, v) \in \neg\Omega(t, v)$, then $\Omega(t, v) \subset \Omega(t, v - 1)$. The iteration procedure of OIS searching continues while condition (5.23) is fulfilled

$$\Omega(t, v) \subset \Omega(t). \tag{5.23}$$

For determination of optimization program OIS set, various D-states $e(t, v)$ formation algorithms can be used. For example, if D-states vectors are binary digits, but D-states are formed adding one unit (see Fig. 5.3), such algorithm is of considerable drawback - technical system graph formation process is occasional and such algorithm can only be used if D-actions number is not large.

Considerably more effective is D-state $e(t, v)$ formation algorithm, which allows performing OIS searching, applying functional characteristics. This kind of algorithm is based on the fact that all D-states set of LTS development optimization tasks can be structured into subsets

$$\{e\} = \{e(t, 0)\} \cup \{e(t, 1)\} \cup \cdots \cup \{e(t, M)\} \cup \cdots \cup \{e(t, n)\}, \tag{5.24}$$

where M realized D-actions number, M is equal to D-state vector χ one unit sum.

In this case, formation of D-state $e(t, v)$ is occurring at $\{e_M\}$, adding a new realized D-action. If such formation method is utilized, then there is opportunity to interrupt D-states formation, if all OIS are determined.

To verify the membership of D-state $e(t, v)$ in the set of OIS, in correspondence with condition (5.20), it is required to calculate functional $f(t, e(t, v))$. This calculation process is labor consuming as transport network with many nodes of production flow distribution must be calculated. Therefore, prior to functional calculation, it must be verified (if it is possible) by fast operable, or even by approximate method, whether $e(t, v)$ can in principle belong to OIS set and such preliminary examination can be logical conditions verification or also control of admissible capital investments.

References

1. Schmock E, Blümel D (1968) Erfahrungen bei der dynamischen Planung elektrischer Üebertragungsnetze auf Digitalrechnern. Energietechnik 18(12):562–565
2. Arzamascev DA, Mizin AL (1970) Choice of thermal power station location and nominal loads by dynamic programming method. In: Применение математических методов и вычислительной техники в энергетике. (Тр. Уральского ордена Трудового КрасногоЗнамени политехн. ин-та им. С. М. Кирова.Сб. 182.) Sverdlovsk, pp 73–81 (in Russian)
3. Costin E (1970) Optimizarea dezvoltăril rețelor electrice urbane prin programare dinamică. Energetica, Bucuresti 18(12):521–526
4. Arie E, Botgros M (1971) Procédés de programmation dynamique et schèmes équivalents employés pour la planification à long terme d'un ensemble de production énergétique. Rev roum sci Techn, Sér 1, 16(3):539–555
5. Pacák S (1971) Dynamická optimalizace při plánováni postupu výstavby elektrických sítí. Energetika 21(11):467–471

6. Kujszczyk S (1971) Optymizacja układów miejskich sieci elektroenergetycznych. Przegl Elektrotechn 47(3):97–101
7. Kujszczyk S (1971) Optymizacja rozwóju osiedlowych sieci elektroenergetycznych metodami programowania dynamicznego. Prace nauk. Politechniki Warszawskiej. Elektryka 24:75–117
8. Dusonchet YP, El-Abiad A (1972) Transmission planning using discrete dynamic optimizing. In: Proceeedings of power systems computation conference, vol 1, Grenoble, p 8
9. Ferenz W, Fiß HJ (1973) Entwicklung und Einsatz der Datenverarbeitung im technischen Bereich eines regionalen Versorgungsunternehmens. Elektrizitaetswirtschaft 72(11):350–360
10. Lee STY, Hicks KL, Hnyilicza E (1974) Transmission expansion by branch- and- bound integer programming with optimal cost-capacity curves. IEEE Trans Power Apparatus Syst PAS-93(5):1390–1400
11. Tröscher H (1973) Entwicklung von technisch-ökonomischen Modellen für die Ausbauplanung. Elektrizitaetswirtschaft 72(15):528–539
12. Arion VD, Zhuravlev VG (1981) Dynamic programming application for electrical energetic system. Штиинца, Kishinev, Moldova (in Russian)
13. Backlund Y, Bubenko J (1978) A computer-aided distribution system planning. Primary substation location and sizing. In: Proceedings of sixth power systems computation conference, vol 1, Guilford, Darmstadt, pp 158–165
14. Cheong HK, Dillon TS (1978) Application of multi-objective optimization methods to the problems of generation expansion planning. In: Proceedings of sixth power systems computation conference, vol 1, Guilford, Darmstadt, pp 3–11
15. El-Abiad AH, Dusonchet YP (1973) Discrete optimization and the planning of electric power networks. IEEE Trans Circuit Theory 20(3):230–238
16. El-Abiad AH, Morin TL, Yamayee ZA (1978) A hybrid dynamic programming branch-and-bound approach to generation planning. In: Modeland simulation proceedings 9th annual Pittsburgh conference, vol 9, Pittsburgh, pp 111–117
17. Fischl R, Schiefele WP (1978) Electric power transmission network planning by nonlinear mixed-integer programming. In: Milwaukee symposium on automatic computation and control, vol 76, Milwaukee, pp 124–130
18. Müller HG, Thieme P, Glimm IA (1978) Optimierungsverfahren zur dynamischen Plannung von elektrischen Übertragungsnetzen. Energietechnik 28(1):9–13
19. Müller HG (1979) Programmsystem zur Optimierung des Ausbaues von elektrischen Versorgungsnetzen. Energietechnick 8:294–298
20. Schmock E (1966) Die Anwendung der dynamischen Programmierung für den optimalen zeitlichen Ausbahn elektrischer Netze. Energietechnik 16(2):65–68
21. Arzamascev DA, Lipes AV, Mizin AL (1976) Models and methods for power system development optimization. Высш.школа, Moscow (in Russian)
22. Benke К, Moloduk VV (1974) Electrical networks development optimization under uncertainty. In: Фактор неопределенности при принятии оптимальных решений вбольших системах энергетики Т. 2. Irkutsk, Russia (in Russian)
23. Venikov VA, Stroev VA (1965) Application of mathematical methods and computers for planning and exploitation of energy systems. Энергия, Moscow-Leningrad, Russia (in Russian)
24. Venikov VA, Zhuravlev VG, Jershevic VV (1972) About software systems for power systems designing. Электроэнергетика и автоматика 14:30–35 (in Russian)
25. Bellman RE (1957) Dynamic programming. Princeton University Press, Princeton
26. Bellman RE, Dreyfus S (1962) Applied dynamic programming. Princeton University Press, Princeton
27. Volkenau IM, Seiliger AN, Habachev LD (1981) Economic of power systems planning. Энергия, Moscow (in Russian)

Chapter 6
Optimal Initial States Searching Methods

Abstract Optimal initial state searching method is essential part of OIS method. It affects on development optimization process duration. Optimal initial state methods with different OIS searching methods together form OIS methods' family. This chapter describes OIS methods searching principles and criteria. It is verified that utilization of LTS properties can greatly improve OIS methods. There is presented LTS functional mathematical model and performed the analysis of the model, which shows the efficiency of model utilization. This chapter describes in detail the analysis of two OIS searching methods principles: System development constraint set searching method and maximal effect method. For these methods OIS searching algorithms and optimization calculation characteristics are given.

6.1 Characteristics of Optimal Initial States Searching Methods

Searching for an OIS set is taking place in the course of functional local maximum determination for D-steps t of specific search area. The above-mentioned corresponds to general optimal initial state definition. In general, search task is an optimal state $\omega(t,i)$ determination, observing condition

$$f(t, \omega(t, i)) = \max_{\{e(t)\}} f(t, e(t)), \tag{6.1}$$

$$\omega(t, j) \subset \neg e(t), \text{where } j = 1, 2, \ldots, i - 1.$$

Maximal functional is calculated using this recursive equation:

$$f(t, e(t)) = Fk(e(t)) + Fr(e(t)) + f(t - 1, \omega(t - 1, m)), \tag{6.2}$$

where $Fk(e(t))$ is the D-state $e(t)$ quality criterion component that is dependent on investments, $Fr(e(t))$ is the D-state $e(t)$ component that is dependent on system reaction to the realized D-actions, $f(t - 1, \omega(t - 1, m))$ is the functional maximal

Z. Krishans et al., *Dynamic Management of Sustainable Development*,
DOI: 10.1007/978-0-85729-062-5_6, © Springer-Verlag London Limited 2011

value that corresponds to system optimal development up to D-state $e(t)$, $(\omega(t - 1,m) \subset e(t))$, and m is the optimal initial state $\omega(t - 1,m)$ index in the data array, arranged.

Functional in a LTS is not common analytical expression; it can only be calculated with numerical methods: D-state quality criterion component $Fk(e(t))$ does not cause any difficulties, but the second component $Fr(e(t))$ calculation is usually labor consuming process. Also determination of functional $f(t - 1,\omega(t - 1,m))$ does not cause any difficulties; optimal initial state $\omega(t - 1,m)$ corresponds to the least OIS index $t - 1$ in D-step, which meets the condition $\omega(t - 1,m) \subset e(t)$. OIS can be determined utilizing recursive equation, if OIS set $\Omega(t - 1)$ and part of the D-step t OIS $\omega(t,j), j = 1,2,\ldots, k - 1$ are known. Putting expression (6.2) into (6.1) results in recursive equation (6.3):

$$f(t, \omega(t,k))$$

$$= \max_{\omega(t-1,m)\in\Omega(t-1)} \left[f(t - 1, \omega(t - 1, m)) + \begin{array}{c} \max \\ \{e(t)\} \\ \omega(t - 1, m) \subset e(t) \\ \omega(t,j) \subset \neg e(t,j) = 1, 2, \ldots, k - 1 \\ \omega(t - 1, i) \subset \neg e(t) i = 1, \ldots, m - 1 \end{array} (Fk(e(t)) + Fr(e(t))) \right].$$

$$(6.3)$$

Adequacy (6.3) is comprehensive expression, where the task is directly formulated to optimize D-state in step t, using the static criterion that must be coordinated with multi-step optimization process, where the integral criterion is utilized. D-state optimization in step t is equivalent to D-state quality criterion

$$F(e(t)) = Fk(e(t)) + Fr(e(t)) \qquad (6.4)$$

maximization, observing optimization area that is represented in expression (6.3).

Application of recursive equation (6.3) is described by examples represented in Fig. 6.1. The set $E = \{e(t)\}$ comprises eight D-states (graph vertices), but D-step $t - 1$ OIS set $\Omega(t - 1)$ contains: $\omega(t - 1,1) = 2$, $\omega(t - 1,2) = 5$, $\omega(t - 1,3) = 1$.

If $k = 1$, then optimal initial state $\omega(t,k = 1)$ is determined as follows. All OIS index values $m = 1$, $m = 2$ and $m = 3$ are considered in sequence.

If $m = 1$, then from set E_{11}, which observes conditions $e(t) \in E$ and $\omega(t - 1,1) \subset e(t)$, the maximal value of $F(e(t))$ and relevant D-state $e(t,v = 1) = 6$ are determined.

If $m = 2$, then from the set E_{12}, which observes conditions $e(t) \in E$, $\omega(t - 1,2) \subset e(t)$ and $\omega(t - 1,2) \subset \neg e(t)$, the maximal value of $F(e(t))$ and relevant D-state $e(t,v = 2) = 5$ are determined.

Fig. 6.1 Example of OIS set determination with recursive Eq. (6.3) for D-states $e(t)$ and $e(j)$, if $e(i) \subset e(j)$. *1* D-states $e(t) \in E(t)$, *2* OIS $\omega(t - 1,m)$ in D-step $t - 1$, *3* D-states $e(t,v)$ with maximal $F(e(t))$, *4* OIS $\omega(t,k)$ in D-step t

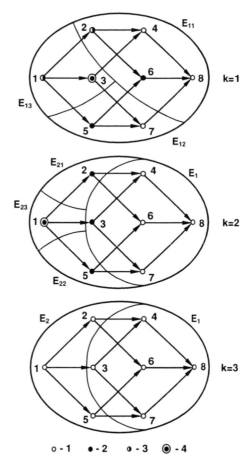

$\circ$ - 1 $\bullet$ - 2 $\circ$ - 3 $\circledcirc$ - 4

If $m = 3$, then from the set E_{13}, which meets conditions $e(t) \in E$, $\omega(t - 1,3) \subset e(t)$, $\omega(t - 1,2) \subset \neg e(t)$ and $\omega(t - 1,1) \subset \neg e(t)$, the maximal value of $F(e(t))$ and relevant D-state $e(t,v = 3) = 3$ are determined.

From the set $\{e(t,v = 1), e(t,v = 2), e(t,v = 3)\}$ according to criterion $\max_{\{1-e}(t,v)\}}[f(t - 1, m) + F(e(t))]$ relevant $e(t) = 3$ is selected. In fact, that is an optimal searched initial state $\omega(t,k = 1) = 3$.

If $k = 2$, then optimal initial state $\omega(t,k = 2)$ searching process is similar; if $k = 2$, additional condition is $\omega(t,1) \subset \neg e(t)$. The result of searching for an optimal initial state is $\omega(t,2) = 1$.

In Fig. 6.1, the set which does not correspond to this condition is E_1. When searching for optimal initial state $\omega(t,2)$, this set is not considered.

In the searched OIS set $\Omega(t)$ there are only two D-states $\Omega(t) = \{3,1\}$, as $E(t,1) \cup E(t,2) = E$. Let us assume that μ and κ are OIS numbers in D-steps $t - 1$ and t, respectively. Then according to recursive equation (6.3) it is required to determine μ and κ local D-states criterion $F(e(t))$ maxima. In the example, considered $\mu \cdot \kappa = 6$.

Fig. 6.2 Graph structure for the system, whose D-states may be maximized by dynamic programming method

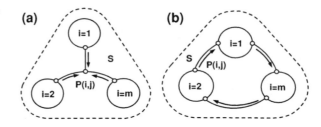

The reviewed principal method can be explicit or approximate. In this case, it is required by explicit method to estimate absolutely all OIS in set $\Omega(t)$. But approximate methods are those which determine only incomplete OIS set. In optimization tasks on LTS development the scope of OIS set (D-states number in total) can be considerable, therefore, approximate OIS methods also have to be used that help to reduce the number of calculable functional local maxima.

Practical algorithms for OIS searching are based on various assumptions on functional characteristics determined by specific task structure. The diversity of real tasks does not allow developing full-scale assumptions on optimal D-process functional characteristics. Therefore, various searching methods are required for OIS identification.

In some optimization tasks, D-state quality criterion $F(e(t))$ can be expressed as a sum:

$$F(e(t)) = \sum_{v=1}^{m} F(t, Y_i).$$ (6.5)

In some cases the task can be formulated so that $F(t(Y_i))$ is only dependent on argument Y_i. In such case in order to calculate the maximal value of $F(e(t))$, in certain conditions dynamic programming can be utilized. DP can be applied for a technical system graph with structures illustrated in Fig. 6.2a, b. The characteristic feature of such structures is the possibility of decomposing the system into m subsystems, which are connected with general configuration network. Dynamic programming can be utilized for subsystems D-states optimization only if the number of network circuits is not higher than one and if argument Y_i can be expressed with dynamic system D-process argument—alternative D-actions set. If these arguments are typical, development dynamic optimization, observing variable discrete and irreversible character, might not be realized.

For systems with multi-circuits networks, the optimization methods are employed, which are based on functional's properties application. Then it is possible to use maximization methods, not considering all D-states. Let us review in detail a model of the functional, which can be utilized in searching process for OIS.

Let us assume that in D-step t any LTS initial graph with multiple circuits is to be considered. System initial graph state is to be marked with $e(0)$.

Fig. 6.3 A mathematical
model of the functional. *1*
Classified realized D-actions
sequence according to functional value augmentation, *2*
Realization sequence
D-actions of occasional
nature

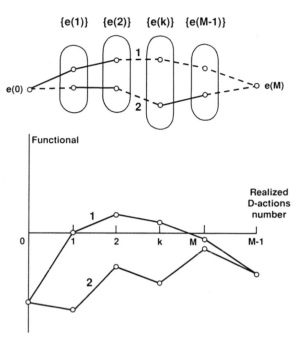

Part of the system D-actions creates one or several new circuits. The system graph, which has *M* new supplementary circuits, will be marked as *e(M)*. In
D-step *t*, new circuits can be subdivided into two groups: effective and ineffective.
Having formed system graph effective circuits, D-state quality criterion $F(e(t))$
value is significantly increased. The formation of ineffective circuits neither
changes $F(e(t))$ value, nor reduces it. The possibility of dividing D-actions into two
groups—effective and ineffective is typical for many real technical system
development optimization tasks. This possibility is based on that in the sophisticated system, spatial interconnections amongst system elements are both rigid and
weak.

Taking into consideration the above-mentioned, we can assume that functional
$f(t,e(M))$ in the system with multiple circuits can be characterized with such
mathematical model:

$$f(t, e(M)) = -\alpha(M_1 + M_2) - \frac{b}{c + M_1} + C_p, \qquad (6.6)$$

where M_1 is effective realized D-actions number, M_2 is ineffective realized
D-actions number, α is the constant that characterizes investment costs in D-step *t*,
b and *c* are the constants that characterizes system response to D-action realization.

Figure 6.3 shows D-states set $\{e_k\}$, where *k* D-actions are realized. Effective
D-actions realization is shown with uninterrupted line, realization of ineffective
D-actions with interrupted line. The functional curve, which is depicted in
Fig. 6.3, will be called D-state graph.

Fig. 6.4 D-actions realization sequence impact on
D-action graph. *1* Realization
sequence is selected at random, *2* Realization occurs
according to criterion
max*Ef*(*e*(*t*,*M*))

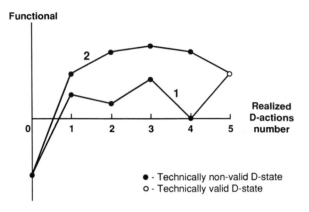

As shown in Fig. 6.3, in D-states graph between D-states $e(0)$ and $e(M)$ various routes from $e(0)$ to $e(M)$ can be found out; each route corresponds to a certain D-actions sequence that alters the technical system graph. If D-actions realization sequence is selected randomly (see Fig. 6.3, curve 2), then there is no regular changes of functional curve. In such cases, effective and ineffective D-actions are realized in turn. Still, D-action realization sequence may be classified in preferable sequence direction if this sequence determination criterion is used:

$$Ef(e(t,M)) = f(t, e(t,M)) - f(t, e(t, M-1)). \tag{6.7}$$

First of all, let us select max$Ef(e(t,M = 1))$, then max$Ef(e(t,M = 2))$ and so on. At the beginning, all D-actions are effective with $Ef(e(t,M)) > 0$, then, as the value of M increases, $Ef(e(t,M))$ value goes down and reaches zero value (maximal functional value), further on $Ef(e(t,M)) < 0$. D-actions with sequence determination criterion $Ef(e(t,M)) > 0$ are effective, but with $Ef(e(t,M)) < 0$, ineffective.

In such a way, selecting D-actions sequence we obtain regular D-state graph, usually with one maximum (see Fig. 6.4, curve 2). Curve 2 in maximum area is very flat.

Functional properties in real technical systems development optimization tasks characterize functional changes in the ordered D-states path.

The results of many experimental calculations allow us to formulate a hypothesis that between any $e(0)$ and $e(M)$ exists at least one realized D-actions sequence, whose D-state functional curve has the shape of ∩ with one maximum.

Moving along determined D-states sequence OIS can be determined [according to definition (6.1)]. OIS correspond to functional local maxima.

It should be noted that in LTS development tasks, functional maximal value achievement is not sufficient for optimal initial state indication. The second requirement is D-state validity with regard to technical limitations, for example, valid flows in network links. There are two methods how to observe technical limitations.

The essence of the first method is that D-states, where technical limitations are not observed, are considered non-valid. If technical limitations in system development task are prevented by this method, then it must be taken into consideration that valid D-states frequently exist in such ∩-shape curve part, where functional value will be reduced. That is represented, for example, in Fig. 6.4. D-states, which have possible maximal functional value and which observe technical limitations, are located behind the boundary between non-valid and valid D-states. Such searching method of D-states will be named constraint state searching method. It will be considered in detail in Chap. 2.

The second method that observes technical limitations is penalty function method. This will be considered in detail in Sect. 6.4, along with optimal states searching maximal effect method.

6.2 System Development Constraint States Set Searching Method

Searching for OIS in LTS development tasks can be performed, searching for ∩-shape functional value curves maxima. Due to that, it is possible not to consider all D-states $\{e(t)\}$, in order to calculate OIS set $\Omega(t)$. The set $\{e(t)\}$ may always be subdivided into two subsets: (1) $\{e(t)'\}$, in which $e(t)'$ can be an optimal initial state, and (2) $\{e(t)''\}$, in which $e(t)''$ is not an optimal initial state. If these subsets are known, to determine $\Omega(t)$, only $\{e(t)'\}$ have to be considered.

Let us review OIS searching, utilizing only the first subset and observing the following conditions:

1. The system represents production transport network with many circuits.
2. One of the reasons for D-action realization is technical limitations violation.
3. The number of D-actions that is realized at the beginning of one D-step is not large (1–5).
4. The number of effective D-actions in one D-step is not large (approximately 10), too.
5. LTS valid D-plan is valid D-states sequence $e(t)$, where t changes within margins $t = 1,2,\ldots, t$. In valid D-state all technical limitations are observed. If in any D-process at least one D-state is non-valid then also D-process is non-valid.

The reviewed LTS D-model is sufficiently adequate to real development circumstances, in which consumption changes in one D-step are relatively not considerable.

Let us formulate major assumptions of OIS searching process observing the above-mentioned stipulations. It can be assumed that in the set of system development graphs there is at least one D-process, in which D-states functional curve is ∩-shaped with one maximum. In electric power systems, it often happens that realized D-actions number that ensures possibility of observing technical limitations is slightly higher than realized D-actions optimal number that corresponds to functional maximal value (see Fig. 6.4).

Fig. 6.5 Constraint searching method. *M* within the
period from $t = 0$ to $t - 1$
realized D-actions number,
R constraint set searching
supplementary realized
D-actions number

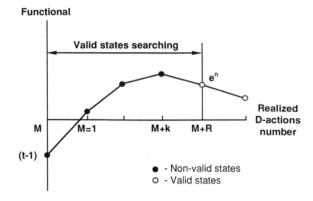

From the aforementioned it follows that if $e(t)'$ is a D-state, where all technical limitations are observed, then D-state $e(t)''$, from which a transition from $e(t)'$ is possible, i.e.,

$$e(t)' \subset e(t)'', \tag{6.8}$$

might not belong to OIS set $e(t)'' \in \neg\Omega(t)$, as

$$f\left(t, e(t)'\right) > f\left(t, e(t)''\right). \tag{6.9}$$

In fact, conditions (6.8) and (6.9) are in contradiction with characteristics of OIS, described in Sect. 5.2. The conditions (6.8) and (6.9) are of the following physical meaning: supplementary D-actions realization does not enlarge system D-state quality criterion value if technical limitations are observed.

OIS searching in D-step t occurs, solving the following task for each valid D-state searching process. For the given $\omega(t - 1)$, it is required to determine all valid D-states $e(t)$, that observe condition

$$\omega(t - 1) \subset e(t)' \subset e(t). \tag{6.10}$$

where $e(t)'$-non-valid D-state.

This task solution is adequate to valid D-state searching process (see Fig. 6.6). In searching, valid D-states set will be referred to as D-states constraint set

$$e(t) \in \Gamma, \tag{6.11}$$

but a method for searching—as *constraint set searching method*.

Let us review realization principles of constraint set searching method. First of all, let us formulate some features of membership in D-states set $\{e''\}$. Let us assume that OIS set $\Omega(t - 1)$ is known. Let us consider OIS $\omega(t - 1, t) \in \Omega(t - 1)$ and form D-states directed graph $G(\omega)$ (see Fig. 6.5). Graph vertices are D-states e:

$$\omega(t - 1, i) \subset e. \tag{6.12}$$

Fig. 6.6 Constraint set
searching directed graph
$G(\omega)$

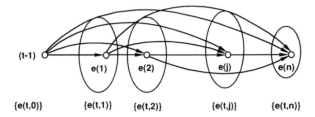

The edges, going out of each vertex demonstrate a possibility of transition from
a concrete vertex to others. Graph $G(\omega)$ edges correspond to definite realized
D-actions set. Graph G edges are named managerial vectors. Managerial vector be
marked with r, but all managerial vectors resulting from summit e, set be marked
with $R(e)$.

All managerial vectors will be divided into two groups: elementary and compound. The elementary vectors correspond to one D-action realization, but the
compound ones correspond to two or more D-actions concurrent realization (in
one D-step). For example, in Fig. 6.6 the vector from $\omega(t-1)$ to $e(t,1)$ is elementary, but the vector from $\omega(t-1)$ to $e(t,2)$ is compound managerial vector.
Elementary managerial vectors sequence will be referred to as managerial path.

Graph $G(\omega)$ vertices (D-states) set is subdivided into subsets that are classified
in the sequential line $\{e(t,0)\},\{e(t,1)\},\{e(t,2)\},\ldots, e(t,j)\},\ldots, \{e(t,n)\}$ (see
Fig. 6.6).

The set $\{e(t,0)\}$ has only one D-state $\omega(t-1)$, but the set $\{e(t,j)\}$ and
$\{e(t,j+1)\}$ vertices are connected only with elementary managerial vectors.

Utilizing the assumed set structure, constraint set Γ is formed by graph $G(\omega)$
nodes, which

1. Are adequate to technically valid D-states;
2. Are connected with initial node $\omega(t-1)$, so are non-valid.

According to functional mathematical model (6.6) between any two valid
vertices $e(t,j)$ and $e(t,j-\alpha)$, included in the path, the following condition is valid

$$f(t, e(t,j)) \leq f(t, e(t, j - \alpha)). \tag{6.13}$$

In order to determine D-states membership in the set $\{e''\}$ let us suggest a
feature. For any vertex $e(t,j)$ of graph $G(\omega)$ it holds that

$$e(t, \Gamma) \subset e(t,j), \tag{6.14}$$

where $e(t,\Gamma)$, D-state that belongs to constraint set, then

$$e(t,j) \in \left\{ e(t)'' \right\}; \quad e(t, j) \in \neg \Omega(t). \tag{6.15}$$

Based on conditions (6.13) and (6.15), we may conclude that only constraint set
Γ D-states can belong to OIS set $\Omega(t)$.

Fig. 6.7 Example of constraint sets D-states searching

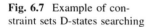
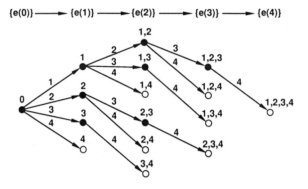

Let us review the process of target constraint sets D-states formation that allows declining D-states $e(t)'' \in \{e(t)''\}$ revision. In this method, initial development states $e(is,t,j-1)$ are formed gradually by D-steps $j = 1,2,..., n$. In each D-step there is the initial states set $\{e(is,t,j-1)\} \subset \{e(t,j)\}$. For each initial state $(is,t,j-1)$, there is valid elementary control vectors set $R(is,t,j-1)$. For elementary managerial vector $r \in R(e,is(t,j-1))$ D-states $e(t,j) = e(is,t,j-1) \cup r$ will be formed. For the new initial state, the following conditions must be observed:

1. D-state is technically non-valid;
2. Valid managerial vectors set $R(t,j)$ must not be empty.

All the formed, technically valid D-states $e(t,j)$ belong to constraint set Γ.

D-states formation process is interrupted if initial state set $\{e(is,t,j-1)\}$ is empty.

In Fig. 6.7, an example of constraint set D-states searching is shown. Here, graph $G(\omega)$ edges are elementary control vectors 1, 2, 3, and 4. Only one edge ends in each graph summit. The formed D-states are signed with used managerial vector numbers.

Let us consider constraint set D-states searching process for the following typical situations:

1. Graph initial state is technically valid. In this case already in the first searching step $j = 1$ an initial states set $\{e(is,t,j-1)\}$ is empty. Initial state belongs to constraint set, but there are no other constrains.
2. Graph initial state is technically non-valid. In this case in searching step $j = 1$ the initial state is $e(is,t,0) = \omega(t-1)$, but valid managerial vectors are 1, 2, 3 and 4. Thus, the D-state set is created $\{e(t,1)\} = \{1,2,3,4\}$. If all set D-states are technically valid, then these belong to constraint set, but initial states set $\{e(is,t,1)\}$ is empty—searching process is interrupted.
3. Graph vertices 1, 2 and 3 are technically non-valid, but vertex 4 is technically valid. In this case, vertex 4 belongs to constraint set, but vertices 1, 2, and 3 form initial states set $\{e(is,t,j=0)\} = \{1,2,3\}$. Searching process can be continued.

Graph vertices correspond to D-state $e(t)$. Technically non-valid D-states are colored in black. Graph edges correspond to managerial (development) actions. The numbers of realized D-actions are ascribed to graph vertices

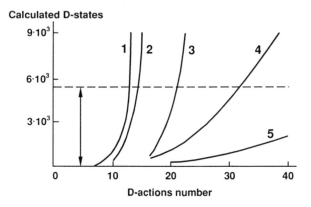

Fig. 6.8 Number of calculated D-states for one optimization step depending on D-actions number. *1* Calculation of all sets $\{e(t)\}$ D-states, *2* constraint set searching having considered also ineffective managerial vectors $n_1 = 0.25n$, *3* calculation of set $\{e(t)\}$ D-states limiting number of managerial vectors $N = 3$, *4* constraint set searching, limiting number of searching steps $N = 3$ and $n_1 = 0.25n$, *5* constraint set searching, excluding ineffective managerial vectors if $n_1 = 0.25n$

Let us compare constraint set D-states searching method with all set $\{e(t)\}$ D-states calculation. All D-states calculation must be performed if characteristics of constraint sets D-states are unknown. In this situation production flow technical scheme calculation model must be estimated (this process is usually labor consuming), as well as other calculations in regard to technical, economic, power supply security, ecological and other criteria; besides, logical conditions must be examined, too.

The total calculation time t_{cal} is calculated by formula

$$t_{cal} = t_1 \cdot V_1 + t_2 \cdot V_2, \tag{6.16}$$

where T_1 is the D-state calculation time, T_2 is the logical operations time, V_1 is the D-states calculation number, and V_2 is the logical operations number. In calculation it can be assumed that $V_1 = V_2 = 2^{10}$.

Figure 6.8 illustrates comparison results of five methods.

As methods comparison criterion, let us use the range of calculated D-states (where system technical, economic etc. calculations have been performed), that is also proportional to calculation time in equal conditions. Still, system calculation time is dependent not only on optimization method, it is also influenced by computer and by LTS model size (number of nodes and edges, number of operational states, model of failure (emergency situation), etc.). If in one D-step about 5,000 D-states must be calculated, then optimization time is too long. The determination of OIS, when calculating entire set $\{e(t)\}$, can be utilized for up to approximately $n = 12$ D-actions. In practical tasks, the number of D-actions is much higher.

In order to reduce the number of D-plans to be calculated, various techniques may be applied, for instance, the method of search steps number limitation, which limits the number of realized D-actions:

$$m(e(t)) \leq m(\omega(t-1)) + N, \tag{6.17}$$

where N be the number of valid supplementary D-actions.

In this method,

$$V_1 = 1 + C_n^1 + C_n^2 + \cdots + C_n^N, \tag{6.18}$$

where C_n^N is the combinations number from n D-actions by N D-actions.

Logical operations number V_2 is calculated by formula

$$V_2 = 2^n.$$

The number of respective calculated D-states V_1 is shown in Fig. 6.8 curve 3. This method can be utilized up to approximately $n = 20$ D-actions.

Constraint set searching method makes it possible to considerably increase the number of alternative D-actions in LTS development optimization tasks. In this method, not all D-states from set $\{e(t)\}$ are considered. In this case

$$V_1 = V_2, \quad V_2 < 2^n.$$

Let us mark total D-state number, obtained in constraint set searching process, with V_1.

In D-step t and concrete optimal initial state $\omega(t-1)$ (if $\omega(t)$ is technically non-valid), all alternative D-actions are subdivided into two groups.

The first group comprises effective D-actions (their total number is n_1), whose realization may fully or partially prevent system technical limitation breach and finally form D-states in which all technical limitations are observed.

The second group comprises ineffective D-actions (their total number is n_2), whose realization does not improve system technical state. It is assumed that all D-actions, which are not included in the first group, are ineffective. The experience demonstrates that n_1 is considerably less than n_2 (approximately $n_1 = 0.25n$).

In the constraint set searching process, illustrated in Fig. 6.6, D-states are formed using effective elementary managerial vector (D-action). In the constraint set searching process, elementary managerial vectors path is formed comprising N effective managerial vectors.

Let us review the following constraint set states searching *algorithm*. In searching steps $j \leq N$ there is formed D-states set sequence $\{e(t,0)\}$, $\{e(t,1)\},\ldots, e(t,j)\}$. All sets comprise only technically non-valid D-states. The respective graph vertices are connected with graph initial state utilizing elementary managerial vectors path, comprising j elementary managerial vectors, out of which $i = 0,1,\ldots, j$ are effective and $j - 1$ are ineffective managerial vectors.

If $j = N$, then all sets $\{e(t,j)\}$ D-states are formed, of which C_n^j D-states are technically valid and are incorporated into constraint set but the rest are initial states for further searching steps.

If $j > N$, only those D-states are formed, which do not correspond to condition (6.15). The vertices of graph G correspond to those D-states, which with initial state connect elementary managerial vectors path. The path comprises not more than $N - 1$ effective managerial vectors. Up to searching step $j = n_2$ (where n_2 is ineffective managerial vectors total number) elementary managerial vectors path might comprise ineffective elementary managerial vector. If $j > n_2$, then the path will contain at least $j - n_2$ effective managerial vectors. Thus, D-states formation process ends in searching step $j = n_2 + N - 1$.

In the reviewed searching algorithm, the number of D-states formed, for which technical, economic and other criteria are calculated, may be determined by formula

$$V_1 = \sum_{j=0}^{N} \sum_{i=0}^{j} C_{n_1}^i \cdot C_{n_2}^{j-i} + \sum_{j=N+1}^{n_2} \sum_{i=0}^{N-1} C_{n_1}^j \cdot C_{n_2}^{j-i}$$
$$+ \sum_{j=n_2+1}^{n_2+N-1} \sum_{i=j-n_2}^{N-1} C_{n_1}^i \cdot C_{n_2}^{j-1}. \tag{6.19}$$

V_1 characteristic curve is depicted in Fig. 6.8, as curve 2.

It is evident that the reviewed algorithm is not effective (many ineffective managerial paths are figured out). Still, this algorithm can be improved limiting the number of search steps up to N. Then V_1 is calculated by (6.19) and only the first item must be considered. Provided that search steps are limited to N, V_1 characteristic curve looks like curve 4 shown in Fig. 6.8, which in comparison with curves 1 and 2, is much flatter. It proves that this method allows increasing D-actions number even up to 30.

Let us review a technique how to more improve constraint set searching algorithm. This approach is based on ineffective elementary searching vector (D-action) excluding from the searching process.

Experimental data obtained out of real systems development optimization tasks, allow figuring out the following elementary searching managerial vectors paths characteristics.

Ineffective managerial vector (D-action) realization forms new D-state in which technical limitations violation is not fully or partly eliminated. Effective managerial vector creates D-state in which technical limitations violation is partly or fully eliminated.

In the formation process of D-states ineffective managerial vectors can be determined which are not utilized in further searching process. In such a way new formed D-states number is considerably reduced. The number of new formed D-states can be calculated by formula

$$V_1 = \sum_{j=1}^{N} C_{n_1}^{j-1}(n - j + 1). \tag{6.20}$$

The characteristic curve of D-states formed with this algorithm is depicted in Fig. 6.8 as curve 5. The curve evidently demonstrates that the algorithm enables solving tasks with sufficiently large D-actions number, which is proved by real development optimization task solution results.

6.3 Constraint Set Searching Algorithms

Constraint set searching process applied in the software is illustrated in Fig. 6.9. D-state 1 is initial state $e(is,t,j-1)$, in which technical limitations are not observed. By adding to initial state managerial vector (realizing D-action 1A), D-state 2 is formed that is also technically non-valid, but, comparing with D-state 1, with decreased technical limitation violation, state 2 in searching step j is preserved as the initial state for searching step $j+1$. Managerial vector 1A is incorporated into effective managerial vectors group. Realizing D-action 2, D-state 3 is obtained, which is technically valid (technical limitation violation are eliminated). D-State 3 is incorporated into constraint set Γ. Also D-action 2A is effective. The next and further searching steps are similar and will continue until all D-states of constraint set are determined.

The principal flowchart of optimization program that has been applied in constraint set Γ searching is shown in Fig. 6.10.

In block 1, initial state $e(is,t,j-1)$ is calculated. This D-state is of previous step $t-1$ optimal initial state $\omega(t-1,w)$. Index number is $w=1,2,\ldots,w_\mathrm{M}$, where w_M is the number of OIS.

In D-state $e(is,t,j-1)$ there are calculated: generation capacity in generation nodes, production flow in transport network, technical, economic and ecological criteria; besides, technical, economic and ecological conditions are verified.

In block 3, constraint set Γ searching for initial state $e(is,t,j-1)$ is carried out, if technical, economic and ecological conditions are not observed thereof. There is only one initial state entering block 3. Constraint set searching occurs gradually realizing D-actions by one up to N. In searching process, new valid D-states are selected. Besides, also new D-states in which limitations violation are reduced only partly are preserved as initial states $e(is,t,j-1)$ of searching process in next step. The number of selected effective initial states is limited (the maximal

Fig. 6.9 Constraint set
searching process

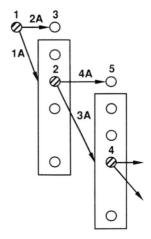

Fig. 6.10 Development
optimization software application for constraint set Γ
searching method. *1* Calculating $e(t,j) = \omega(t - 1,w)$, *2*
$\Gamma = 0$, *3.1* $\{e''(t,j)\} = 0$,
3.2.1 to state $e(t,j - 1)$ parallel paths are connected,
3.2.2 forming Γ, *3.2.3* calculating $\Delta\pi_{min}^{*} = \min \Delta\pi_{i}^{*}$,
3.2.4.1 calculating value $\Delta\pi_{i}^{*}$,
3.2.4.2 eliminating circuit I,
3.2.4.3 calculating state
$e(t,j - 1,i,l)$, *3.2.4.4* forming
Γ, *3.2.4.5* forming set
$\{e''(t,j)\}$, *4* forming $\Omega(t)$

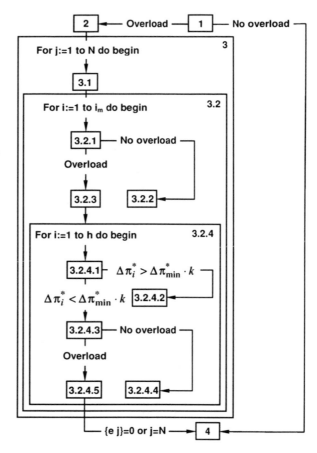

selected initial states number is i_M). Sets $\{e(is,t,j - 1)\}$ and Γ formation is taking
place for all initial states $e(is,t,j - 1,i)$, reviewing $i = 1,2,..., i_M$.

In block 3.2.1 to all initial states $e(is,t,j - 1,i)$ with overloaded transport network links parallel links are connected. In fact, this kind of approach creates new
D-actions, which have not been envisaged drawing up optimization task. For new
created D-state flows, technical, economic and ecological criteria are calculated. If
the new created D-state is technically valid, it is included in constraint set Γ.

In block 3.2.3 an analysis of transport network is performed—it is verified how
the connected parallel links influence the initial overload of state $e(is,t,j - 1,i)$
links.

In block 3.2.4, connecting to transport network state $e(is,t,j - 1,i)$ new circuits
$l = 1,2,..., h$, initial states are formed $e(t,j - 1,i,l)$ and their efficiency is verified
with high-speed operable, though approximate methods. If there is no such method
in the concrete LTS, then block 3.2 is excluded from software flowchart. In
development optimization of electric power system the perfusion methods and
mutual replacement principles can be utilized [1].

Constraint set searching process is finished if at least one of the following conditions is fulfilled:

1. Initial state $e(is,t,j)$ set is empty;
2. The number $j > N$ of valid searching steps is exceeded;
3. In block 4, from constraint set Γ states there is formed an OIS $\omega(t)$ set $\Omega(t)$. Formation algorithm was illustrated in Sect. 5.3.

Figure 6.11 presents experimental data which describe electric power system development optimization program that employs perfusion methods and mutual replacement principles in transmission network estimation.

Program searching process efficiency is estimated by the following criterion

$$\Delta \pi^*(l) \leq \Delta \pi^*_{min} \cdot k. \tag{6.21}$$

where $\Delta\pi^*(l)$ is the load flow relative deviations in the most loaded network links of new circuit l resulting from integrity, $\Delta\pi^*_{min}$ is the load flow deviations in the most loaded network links in the most effective circuit l as a result of integrity, k is the constant $0 < k < 1$ that determines searching process sensitivity

For a specific task of concrete electric power system development optimization, the following constraint set searching process characteristic values are taken into account:

1. The total number V_2 for D-states formed in constraint set searching process. This value is determined by optimization task scope (nodes and links number, D-steps number, D-actions number) as well as by valid searching steps number N. If for determination of effective actions, quick-operating approximate methods are not applied, then V_2 is equal to the number of D-state technical, economic and ecological criteria calculations, V_1.

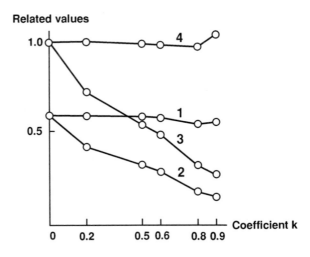

Fig. 6.11 Characteristic curves of constraint sets searching process for various coefficient k values. *1* Selectable effective D-states number V_3/V_2, *2* selected D-states number V_4/V_2, *3* calculation time saving V_4/V_3, *4* functional values changes

2. D-states number V_3. Such D-states are formed by connecting effective circuits. V_3 corresponds to calculation number, if $k = 0$. In such a case in constraint set searching all ineffective circuits are not deselected.
3. The number of D-states V_4, in which new circuit connection caused partial relief of overloaded links according to (6.21). Ratio V_4/V_2 denotes the number of those D-states calculations, which have been performed in constraint set searching process period. As shown in Fig. 6.11, this ratio may be altered by changing value k. Ratio V_4/V_2 demonstrates the application efficiency of D-actions quick-operating approximate estimation methods.

6.4 Maximal Effect Method

Functional $f(t,e)$ changes in LTSs can be shown in a way that allows one to use *the maximal effect method* for searching OIS set. As the D-state is a set of realized D-actions, $\{e(t)\}$ can always be expressed by subset sequence $\{e(0)\}$, $\{e(1)\},\ldots, \{e(M)\},\ldots, \{e(n)\}$, where M is the number of realized D-actions in D-state $e(M) \in \{e(M)\}$. If the D-state is described with the binary vector χ, then M is the sum of components values (the number of components, whose value is equal to 1). For D-state $e(t,M)$ the functional looks like this

$$f(t, e(t, M)) = F(e(t, M)) + f(t - 1, \omega(t - 1)), \qquad (6.22)$$

where $\omega(t - 1)$ is the optimal initial state $\omega(t - 1) \in \Omega(t - 1)$ with maximal $f(t - 1, \omega(t - 1))$, if $\omega(t - 1) \subset e(t,M)$.

D-state quality criterion $F(e(t,M))$ may be structured into two components:

$$F(e(t, M)) = Fk(e(t, M)) + Fr(e(t, M)), \qquad (6.23)$$

where $Fk(e(t,M))$ is the component of quality criteria state that is dependent on capital investments in realized D-actions, $Fr(e(t,M))$ is the component of quality criteria state that reflects system response to realized D-actions

For each D-state $e(t,M)$, D-states sequence (path) exists $e(0) \subset e(t,1) \subset e(t,2) \subset \cdots \subset e(t,M - 1) \subset e(t,M)$; there is at least one such path. In general case there are $M(M - 1),\ldots, 3,2,1$ paths. Each path can be described by *the effect function.*

$$Ef(M) = f(t, e(t, M)) - f(t, e(t, M - 1)). \qquad (6.24)$$

If $Ef(M) > 0$, D-action M is effective, its realization gives a positive effect. If $Ef(M) < 0$, D-action M is not effective, its realization produces a negative effect.

In Fig. 6.12, effect function is shown for D-states path of random character.

As can be seen from Fig. 6.12, for random path (if $M = 5$, then such paths number is $V_k = 1,2,3,4,5 = 120$) function $Ef(M)$ curve A is fluctuant and its changes are regular. Such a function cannot be used in optimal initial state

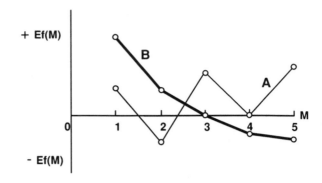

Fig. 6.12 Maximal effect function. *A* case characteristics for path, *B* for target coordinated path

searching process. Our target is to create OIS searching methods to figure out a tree of OIS in searching process (to arrange an optimal D-states path). Then effect function possesses a regular form, it may look, for example, like curve B (see Fig. 6.12). In such case, it is easy to determine all OIS and also to determine the final stage of searching process.

As shown in Fig. 6.12, up to $M = 3$ (the maximal function value) all $Ef(M)$ are larger than 0, but, if $M > 3$, then $Ef(M)$ are less than 0. Searching can be interrupted if $M = 3$, because as our analytical and experimental research data show, the ordered path functional $f(t,M)$ curve has only one maximum where $Ef(M) = 0$.

Let us mark searching OIS in step t with $\omega(t,i,M)$. Value i denotes optimal initial state index, in the sequence arranged to the functional values. Amplifying i value, functional value is decreasing. Value M denotes the number of realized D-actions. In order to be able to apply the maximal effect method to searching for OIS, let us move from graph vertices $e(0)$ along OIS trees branches. First, let us determine OIS tree.

OIS nodes of a tree are OIS $\omega(t,i,M)$, but branches connect nodes $\omega(t,i,M-1)$ with $\omega(t,k,M)$, if

$$\omega(t,l,M-1) \subset \omega(t,k,M). \qquad (6.25)$$

If in searching step $M - 1$ there are several D-states, from which the transition to optimal initial state $\omega(t,k,M)$ is possible, then it is necessary to select tree branch with the highest value of the maximal effect function.

$$\max Ef(M) = \max_{\omega(t,l,M-1)\subset\omega(t,k,M)} [f(t,k,M) - f(t,l,M-1)]. \qquad (6.26)$$

$$\max Ef(M) > 0. \qquad (6.27)$$

It means that the most effective D-actions must be applied in this case. The rest graph edges are not contained in the tree. Therefore, no loops of the tree are developed.

Since all tree branches are max$Ef(M) > 0$, then, if max$Ef(M + 1) < 0$, the respective graph edge is out of the tree and searching for optimal initial state in this direction is interrupted.

Fig. 6.13 Functional mathematical model for OIS
searching applying the maximal effect method. *1* functional $f(t,e(t,M))$, disregarding
technical limitations, *2* functional $fp(t,e(t,M))$, taking into
consideration penalty function for technical limitation
violation, M_{opt} number of
optimal realized D-actions,
M_r the number of D-actions
that prevent technical limitations violation

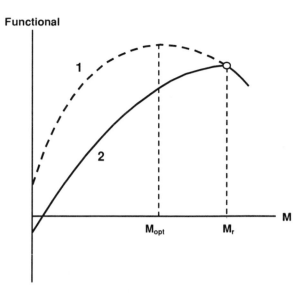

The main characteristics of OIS tree are as follows:

1. Development initial state $e(t,0)$ is always contained in the OIS set; $e(t,0)$ functional has the smallest value, thus $\omega(t,i,M = 0)$ index i is of the highest value, $e(t,0)$ is always placed in the end of the ordered line of OIS;
2. All D-states of OIS tree are OIS $\omega(t,i,M)$. Any optimal initial state can be connected with OIS tree initial state only with one path: $\omega(t,i,0) \subset \omega(t,i,1) \subset \cdots \subset \omega(t,i,M) \subset \cdots \subset \omega(t,i,M + k)$. The higher is the realized D-action number M, the greater the value of functional and the lesser the value of index i;
3. D-state $\omega(t,M)$ is an optimal initial state if in the whole path, in which $e(t,M)$ is included, $\max Ef(N) > 0$ for all N values, $N = 1,2,\ldots, N,\ldots, M - 1, M$.

The hypothetical character of functional changes—with one maximum (or no one, the maximum is located prior to initial D-state $e(0)$)—corresponds to the real systems with sophisticated production transport network when D-action M realization alters production power balance distribution in the transport network.

When applying the maximal effect method, technical limitations violation must be estimated by penalty functions. The method is described in Fig. 6.13, where M_r, realized D-actions, which are required for preventing technical limitations violation. If $M < M_r$, then technical limitations are not taken into consideration. Curve 1 shows changes in functional $f(t,e(t,M))$ depending on the number of realized D-actions. Functional value is maximal, if the number of realized D-actions is optimal M_{opt}. If $M_r > M_{opt}$, then the OIS obtained with the maximal effect method will be technically non-valid D-states.

If technical limitations violation is estimated by penalty function $Fp(t,e(t,M))$, a new functional can be used:

$$fp(t, e(t, M)) = f(t, e(t, M)) + Fp(t, e(t, M)), \tag{6.28}$$

where $Fp(t,e(t,M))$ is the penalty function.

Penalty function is estimated concurrently with D-states technical criteria calculation(see Sect. 1.2).

For technically valid D-states $Fp(t,e(t,M)) = 0$.

A penalty function must be selected (generally by simulation), so as to ensure determination of technically valid OIS (in principle, similar to the determination of constraint sets). If new functional value changes depending on the realized D-action similar to curve 2 (see Fig. 6.12), then the maximal effect method ensures determination of constraint set.

As approved by experience in pilot large power engineering systems development optimization tasks estimation, formula (6.28) ensures application of the maximal effect method.

Let us review an example, optimization of integrated 330 kV transmission network. Network model incorporates 51 nodes and 91 links; 13 different D-actions are realized. In the network considered, all OIS are determined.

The properties of the previously described functional $f(t,e(t,M))$ are valid for the expanded area of development optimization tasks (determined by system development technical and economic indicators). As the number of realized D-actions M is increasing, absolute values of the component $Fk(e(t,M))$ of D-state quality criterion are increasing too, but D-state quality criterion component that reflects system response to the realized D-actions, $Fr(e(t,M))$, is decreasing. Therefore, by increasing the number of realized D-actions M, the curves of development criterion $C–F,e(t,M)$ (where C revenues due to sold products and $F,e(t,M)$, quality criterion of D-state $e(t,M)$ in step t), become ∩-shaped.

Figure 6.14 explains one of possible algorithms for optimal initial state searching OIS with the maximal effect method. The flowchart of the algorithm is shown in Fig. 6.15.

The searching process of OIS tree nodes occurs gradually, in augmentation direction of realized D-actions number M. At the level M of OIS searching process, it is necessary to review C_n^M different D-states $e(t,M)$ (where n total number of D-actions). The D-states shall be estimated in the following sequence: (1) a, at searching level $M = 0$ in D-state $\{e(t,0)\}$; (2) at searching level $M = 1$ in D-states b, c, d, e, $\{e(t,1)\}$, and so on.

According to the criterion $\max Ef(e(t,M))$ values signs (positive/negative), we determine whether the D-states $e(t,M)$ belong to OIS set $\Omega(t)$, and determine further character of functional changes. As shown in Fig. 6.14, OIS are estimated as follows: an initial state $e(t,0)$; D-states e, d, and b forming a set $\{e(t,1)\}$; D-state h is incorporated in set $\{e(t,2)\}$. Such D-states as k are not OIS, as $\max Ef(t,k) = f(t,k) - f(t,d)) < 0$.

The calculation of the maximal effect function is the most labor and time consuming process, if a network has a large number of circuits, equations systems with many Y, X, Z must be solved. The maximal effect method is effective only if greater part of D-states $e(t)$ can be eliminated without calculating the maximal effect function $\max Ef(e(t))$. We can use the following D-state $e(t,M)$ non-membership indication in OIS set $\Omega(t)$. If

Fig. 6.14 OIS set searching with the maximal effect method. *1* OIS, *2* D-states, whose membership in the set of OIS set must be examined by max*Ef*(*e*(*t*,*M*)) criterion, *3* D-states, for which max*Ef*(*M*) must not be calculated, *4* D-states, in which searching processes shall be interrupted, *5* edges with max*Ef*(*M*) > 0 *6* edges with max*Ef*(*M*) < 0, *7* OIS tree branches

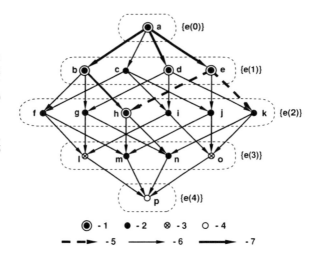

$$\omega(t, M-1, k) \subset \neg e(t, M), \quad k = 1, 2, \ldots, \tag{6.29}$$

then $e(t,M)$ has no OIS tree nodes. In Fig. 6.14, such D-states are D-states l and o, as not all nodes of paths, which connect the initial node a to nodes o and l, are OIS.

If state non-membership in OIS set $\Omega(t)$ is expressed as

$$\omega(t, M-\alpha, k) \subset \neg e(t, M), \quad \alpha > 1, \quad k = 1, 2, \ldots, \tag{6.30}$$

then even such local functional maxima will be found where $f(t,M)$ deviations occur (usually in optimum zone which is flat). Still, this solution specification occurs at the expense of calculation time extension. The value of coefficient α regulates optimization algorithm accuracy and operation speed.

If in the set $\{e(t,M-1)\}$ no optimal initial state is determined, then OIS will not be in the set $\{e(t,M)\}$. Due to chat, searching for OIS is interrupted.

The reviewed method enables determination of all OIS. First, initial states with less functional values are determined (for these D-states indexes i will be higher). When increasing the number of realized D-actions M, indexes i are decreasing. The latter is essentially important for the practical application of this method, because less local maxima of the functional must be determined.

Let us estimate the valid scope (the number of D-actions) for the development optimization task, which can be solved with the maximal effect method. Let us express functional with idealized mathematical model (6.6). If $M = M_{opt}$, then functional is of maximal value. In mathematical model step t, all D-actions n are divided into two groups: effective D-actions n_1—D-actions, which amplify functional value, and ineffective D-actions n_2—D-actions that reduce functional value. OIS are a combination of realized effective D-actions.

OIS searching process in D-step (level) M will be specified by the following values:

1. The number of created D-states

$$B(M) = C_n^M, \tag{6.31}$$

Fig. 6.15 Searching for OIS set with the maximal effect method algorithm. *1* Create a D-state $e(t,M,i)$, *2* first examination $e(t,M,i) \in \neg\Omega(t)$, *3* verify logical conditions, *4* calculate $f(t,e(t,M,i))$, $(t,e(t,M,i))$, *5* verify $e(t,M,i) \in \Omega(t)$, *6* alter $\Omega(t) = \Omega'(t) \cup e(t,M,i)$, *7* whether all $e(t,M,i)$ are reviewed, *8* or $\Omega(t,M) = \Omega(t,M-1)$

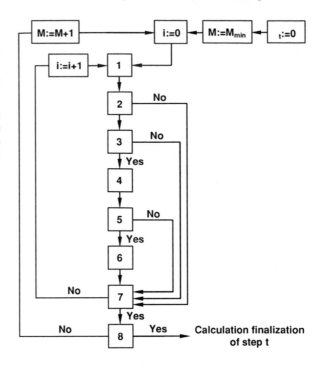

2. The number of optimal states

$$W(M) = C_{n_1}^M, \tag{6.32}$$

3. The number of D-states for which the maximal effect function $\max Ef(e(t,M)$ must be calculated

$$H(M) = C_{n_1}^M + C_{n_1}^{M-1} \cdot n_2. \tag{6.33}$$

The corresponding characteristic curves are depicted in Fig. 6.16.

In order to characterize OIS searching process efficiency, let us apply the criterion *relative efficiency*

$$\sum_{M=0}^{n} C_n^M \bigg/ \sum_{M=0}^{n} H(M). \tag{6.34}$$

In formula (6.33), fraction numerator corresponds to the area below curve 1, but denominator, to unshaded area below curve 2. The relatively by lesser, n_1/n and M_{opt}/n, the higher the relative efficiency of the proposed method.

Figure 6.17 shows the number of efficient D-actions as a function of D-actions number for real electric power systems development optimization tasks. To solve this task, the maximal effect method is applied.

Fig. 6.16 Characteristic curves of optimization D-states searching process if $n = 11$, $n_1 = 8$, $M_{opt} = 4$. *1* Number of created D-states $B(M)$, *2* number $H(M)$ of D-states for which the maximal effect function max-$Ef(e(t,M)$ must be calculated, *3* number of OIS $W(M)$

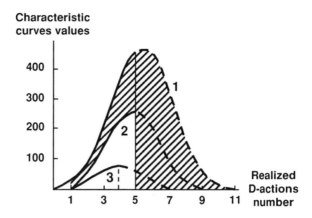

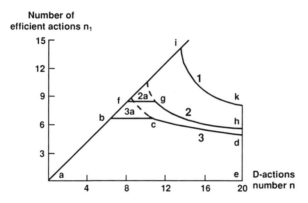

Fig. 6.17 Characteristic curves of OIS searching with the maximal effect method maxEf. *1–3*—the highest valid values of n and n_1, if $M_{opt} = 2,3,5$, not observing limitations of computer main memory; *2a* and *3a*—the highest valid n and n_1 values, if $M_{opt} = 3$ and 5, observing limitations of computer main memory *us* (it is assumed that up to 100 OIS may be stored); *a, b, c, d, e; a,f, g, h, e; a, b, i, k,* and *e*—valid n and n_1 values, if $M_{opt} = 5, 3, 2$

If computer RAM capacity is not sufficient to store all OIS, then the limited length data array can be used, as it usually does not produce any errors in optimization results.

The analysis of practical application of the maximal effect method evidently demonstrates the capability of the method to rapidly solve the tasks on development dynamic optimization.

Reference

1. Dale VA, Krishans ZP, Paegle OG (1979) Dynamic methods for analysis of power system networks development. Zinatne, Riga (in Russian)

Chapter 7
LTS Sustainable Development Management Approximate Methods

Abstract This chapter presents approximate LTS development optimization methods: (1) optimal initial state number limitation method, (2) limitation method of OIS searching process, (3) dynamic development tasks solution iteration method, and (4) optimization method in adaptation period. All the methods presented are based on optimal initial state dynamic optimization methods. Software of the method can be developed, improving software of the methods, which are reviewed in Chaps. 5 and 6. For all methods, an analysis of calculation aspects is performed, and respective characteristics are defined. The results have shown that approximate methods' utilization can considerably enlarge the number of alternative development actions. In majority of real tasks, it gives opportunity to get more optimal development process plans as compared to exact method utilization.

7.1 Approximate Optimization Methods for Development Management

Usually optimization methods are approximate. Their approximate nature is determined not only by calculation methods objective function but mainly by approximate methods of optimal D-plan searching. Dynamic method inaccuracy is caused by the fact that not all competitive states are preserved in optimization steps. As a result, there is a certain probability that global optimum will not be determined. The accuracy of the approximate method may be characterized by the difference between best objective function value of approximate D-plan and global optimum objective function (that is usually difficult to be identified) value. However, the application of approximate optimization methods in no way means that the made decisions are aggravating. On the contrary, just because the variables number is increased, the obtained solutions have better technical and economic indicators.

Approximate optimization methods are a logical sequel of dynamic optimization methods (including also OIS methods) that are reviewed in Chaps. 2–6.

Z. Krishans et al., *Dynamic Management of Sustainable Development*,
DOI: 10.1007/978-0-85729-062-5_7, © Springer-Verlag London Limited 2011

Dynamic optimization methods reviewed in the mentioned chapters serve as the basis for approximate dynamic optimization methods.

In Sect. 7.2, there are two methods analyzed: OIS set limitation method and OIS searching process limitation method. In the elaboration of these methods, equally significant was the following: theoretical analysis, experimental calculations, and methods' utilization practical analysis. All three aspects are discussed in this chapter.

7.2 Optimal Initial States Number Limitation Methods

With an increase in D-actions number (variables), OIS number $W(M)$ also increases rapidly. This can be estimated by the following formula:

$$W(M) = \sum_{M=0}^{M_{opt}} C_{n_1}^M, \qquad (7.1)$$

where n_1 effective D-actions number; M realized D-actions number; M_{opt} realized D-actions number, to which the maximal functional value corresponds

If $n_1 = 10$ and $M_{opt} = 5$, then OIS number $H(M)$ is as large as 600.

D-states technical, economic and ecological criteria calculation number $H(M)$ is:

$$H(M) = \sum_{M=0}^{M_{opt}} C_{n_1}^M + (n - n_1) \cdot \sum_{M=0}^{M_{opt}} C_{n_1}^M. \qquad (7.2)$$

If $n = 15$, $n_1 = 10$, and $M_{opt} = 5$, then OIS number $H(M)$ is already 2,600.

If OIS number $H(M) = 2,600$, then, in using of OIS method with all OIS storing, certain problems might occur.

7.2.1 Optimal Initial States Set Limitation Method

There is a possibility that only part of OIS are stored, at the same time not considerably reducing objective function value as well as not altering optimal D-process in principle.

Practical experience of OIS methods application shows that the probability of optimization result variation is slight if $n = 10–12$ and if in steps $t - 1$ and t only 30–50 best OIS are stored.

By reducing the OIS number $W(M)$, stored optimization process can be considerably accelerated. This real fact is evidently demonstrated by experimental calculation results (see Fig. 7.1).

The task on particular electric power system development optimization is solved with 6 D-steps and 16 D-actions. As shown in Fig. 7.1, as the number of OIS increases, the calculated OIS number increases almost linearly. However,

Fig. 7.1 OIS calculation number dependency on OIS number

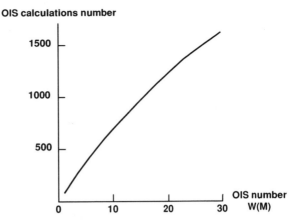

optimization results in the reviewed experiment begin to change visibly if $W(M) < 5$.

Let us analyze the problem occurring due to the reduction of OIS number. Figure 7.2 presents analytical research results of functional value distribution of the OIS set.

Curve 1 demonstrates how functional relative value is changing. The maximal value (item a) corresponds to such optimal initial state that has M_{opt} realized D-actions. States with $M > M_{opt}$ realized D-actions are not OIS. Indeed, to the right from item a, functional value is reducing. Therefore, for any state, $e(t, M > M_{opt})$ always be found such optimal initial state $\omega(t)$, which observes the conditions $f(t, e(t, M)) < f(t, \omega(t))$ and $\omega(t) \in e(t, M)$. It means that $e(t, M > M_{opt})$ is not an optimal initial state.

Curve 2 shows OIS number $W(M)$ relative characteristic curve. OIS set is complete if $W(M) = 1$. OIS number in complete set can be calculated by Eq. 7.2. The value $\Delta f''$ characterizes functional changes range in the limited OIS set.

Relative characteristic curves may be utilized to estimate optimization process for concrete D-actions number. If functional is changing in the range $\Delta f''$, OIS number can be calculated as follows:

$$W(M2) = W(M1) \cdot 2^{n_1' - n_1} \cdot W^*(M), \qquad (7.3)$$

where $W(M1)$ OIS number, if D-actions number is n_1; $W(M2)$ OIS number in the limited set, in which functional changes in the range $\Delta f''$; $W^*(M)$ relative D-states number in the limited set; $n'1$ effective D-actions number.

7.2.2 Limitation Method of Optimal Initial States Searching Process

Let us review approximate system development optimization method in which OIS are searched by applying the maximal effect method. The search for OIS results in

Fig. 7.2 OIS set relative characteristic curves. *1* functional relative values; *2* OIS integral number dependence on realized D-actions number; *3* maximal effect function max $Ef(e(t, M))$ relative values; *A* OIS with max $Ef(e(t, M)) \leq \delta$ number; *B* OIS with max $Ef(e(t, M)) > \delta$ number; $\Delta f''$ functional changes for best OIS, if max $Ef(e(t, M)) \leq \delta$

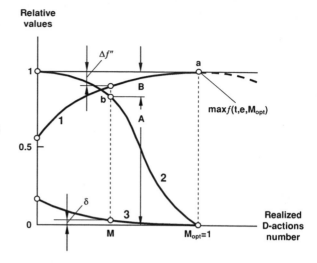

the creation of a path of states. Path states functionals are changing in logical regular sequence (see Fig. 7.3).

In order to describe the new approximate development optimization method, let us utilize another kind of definition of OIS tree branches.

Tree branches are transitions from state $\omega(t, M - 1, i)$ to state $\omega(t, M, k)$ that observe conditions:

$$\max Ef(t, M, k) = \max_{\omega(t, M-1, i) \subset \omega(t, M, k)} [f(t, \omega(t, M, k)) - f(t, \omega(t, M - 1, i))], \quad (7.4)$$

$$\max Ef(t, M, k) > \delta. \qquad (7.5)$$

Comparing Eqs. 7.4 and 7.5 with Eqs. 6.26 and 6.27, it can be seen that there are changes in the second condition (Eq. 7.5). Now it corresponds to approximate

Fig. 7.3 Functional $f(t, e(t, M))$ changes in OIS tree: an example on real electric power system development optimization

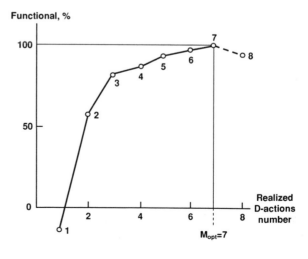

functional maximal value calculation. The essence of the method is in the point that only the most effective D-actions are used here.

Let us review method properties in solving system development optimization tasks. Let us assume that in optimization process two principally different development strategies comparison is fulfilled. Each D-action, both a and b correspond to their own system development strategies. In step t, D-actions a and b realizations create the highest functional value augmentation. Let us assume that there are also other D-actions: i, j, etc. in the task. This D-actions realization produces considerably smaller functional value augmentation. If coefficient (admissible smaller max $Ef(e(t))$ value) $\delta = 0$, then it might occur that D-actions combination a, i, j and b, i, j could not be stored. In such a case, there is a possibility that task dynamic meaning will be missed. If coefficient $\delta > 0$, then task result is a strategy with D-actions a and b, which are considered and stored within the entire estimation time period.

If coefficient $\delta > 0$, then OIS searching process characteristic curves can be estimated using characteristic curves depicted in Fig. 7.2. Curve 3 demonstrates that with an increase in the realized D-actions number, max $Ef(e(t, M, k))$ is changing as well. Let us assume that OIS are divided into two groups: max $Ef(e(t,$

Table 7.1 System development optimization task results

$\delta(\%)$	t	\multicolumn{16}{c}{D-state binary vector for optimal D-plan}	$f(t)$
---	---	1	2	3	4	5	6	7	8	9	10	11	12	13	14	15	16	
0	1	1	0	0	0	0	0	0	0	0	0	0	0	0	0	0	0	1.000
	2	1	0	1	0	0	0	0	0	0	1	0	0	0	0	0	0	1.000
	3	1	1	1	0	0	0	0	0	0	1	0	0	0	0	0	0	1.000
	4	1	1	1	0	0	0	0	0	0	1	0	0	0	0	0	0	1.000
	5	1	1	1	0	0	0	0	0	0	1	0	0	0	0	0	0	1.000
	6	1	1	1	1	1	0	1	0	0	1	0	0	1	0	1	1	1.000
0.5	1	1	0	0	0	0	0	0	0	0	0	0	0	0	0	0	0	1.000
	2	1	0	1	0	0	0	0	0	0	1	0	0	0	0	0	0	0.998
	3	1	1	1	0	0	0	0	0	0	0	0	0	0	0	0	0	0.986
	4	1	1	1	0	0	0	0	0	0	1	0	0	0	0	0	0	0.994
	5	1	1	1	0	0	0	0	0	0	1	0	0	1	0	0	0	0.984
	6	1	1	1	0	0	0	0	0	0	1	0	0	1	0	0	0	0.984
1	1	1	0	0	0	0	0	0	0	0	0	0	0	0	0	0	0	1.000
	2	1	0	1	0	0	0	0	0	0	1	0	0	0	0	0	0	0.998
	3	1	1	1	0	0	0	0	0	0	0	0	0	0	0	0	0	0.982
	4	1	1	1	0	0	0	0	0	0	0	0	0	0	0	0	0	0.980
	5	1	1	1	0	0	0	0	0	0	0	0	0	0	0	0	0	0.970
	6	1	1	1	0	0	0	0	0	0	0	0	0	0	0	0	0	0.954
3	1	1	0	0	0	0	0	0	0	0	0	0	0	0	0	0	0	1.000
	2	1	0	1	0	0	0	0	0	0	1	0	0	0	0	0	0	0.998
	3	1	0	1	0	0	0	0	0	0	1	0	0	0	0	0	0	0.956
	4	1	0	1	0	0	0	0	0	0	1	0	0	0	0	0	0	0.910
	5	1	0	1	0	0	0	0	0	0	1	0	0	0	0	0	0	0.900
	6	1	0	1	0	0	0	0	0	0	1	0	0	0	0	0	0	0.860

M)) $\leq \delta$ (Fig. 7.2, line segment B) and max $Ef(e(t, M)) > \delta$ (Fig. 7.2, line segment A). Most of OIS have non-high max $Ef(e(t, M))$ value, for instance, approximately, for 80% of OIS max $Ef(e(t, M)) \leq 0.003$. If coefficient $\delta = 0.03$, then OIS number is only 20% of all complete set states. In this case, the maximal functional value is aggravating only by 0.1, compared to optimization results if coefficient is $\delta = 0$.

Table 7.1 demonstrates results of electric power system optimization task. There are 6 D-steps and 16 alternative D-actions in the task. If coefficient $\delta > 0$, then optimization results are approximate: the maximal functional value in the least optimization steps is less than 1. Still, if $\delta = 0.005-0.010$, functional error in the least optimization steps is not high (1–5%). With such objective functions' distinctions, the D-plans can be considered equivalent, particularly if taking into consideration forecast credibility for a 20–30-year perspective.

Table 7.1 shows that coefficient δ value may be used as D-action effectiveness criterion. The highest effectiveness is of D-actions 1, 2, and 3. If coefficient $\delta = 0.000, 0.005,$ and 0.010, then these D-actions are realized. The rest D-actions realization expediency is dependent on coefficient δ value. Combinations of many effective and less effective D-actions are created which, according to objective function value, are equal. By increasing coefficient δ value, such combinations are deselected.

The reviewed approximate OIS searching method considerably accelerates system development optimization process as it is illustrated in Fig. 7.4. By increasing coefficient δ value, optimization time is decreasing because M_{opt} value is decreasing as well. The consideration is significant too that OIS may be stored in a relatively much wider range. In such a way, strategically significant optimal states required for dynamic optimization tasks solution are selected.

Experience analysis of electric power systems' development optimization estimation gives evidence that as coefficient δ value increases, functional characteristic

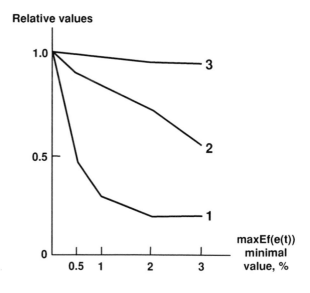

Fig. 7.4 Relative characteristic curves for the approximate OIS searching method. *1* system D-states calculation number $H(M)$; *2* optimal realized D-actions number M_{opt}; *3* functional maximal value

curves become sharper—they become equivalent to functional characteristic curves, obtained if optimization task with fewer alternative D-actions is solved. Concurrent application of the maximal effect function max $Ef(e(t))$ minimal value limitation method and preserving OIS set limitation method application is also possible. By combining two methods, we can increase D-action number even more.

Finally, let us emphasize one more structural property of the optimization process of the maximal effect function max $Ef(e(t))$ minimal value limitation method. If optimization is performed with coefficients δ and δ', where $\delta' > \delta$, then we obtain these states sequences for optimal D-plans:

$$e(0) \subset e(1,\delta) \subset \ldots \subset e(t,\delta) \subset \ldots \subset e(T,\delta),$$
$$e(0) \subset e(1,\delta') \subset \ldots \subset e(t,\delta') \subset \ldots \subset e(T,\delta').$$

Optimization process structural property is that for any D-step t the following condition is valid:

$$e(t,\delta') \subset e(t,\delta), \tag{7.6}$$

which can be verified by Table 7.1 data. Equation 7.6 establishes a theoretical basis for solving system development optimization tasks by iteration methods, gradually reducing coefficient δ values. Such method is discussed in detail in Sect. 7.3.

7.3 Dynamic Development Tasks Solution Iteration Methods

7.3.1 Direct and Iteration Methods

Direct methods examine all alternative D-actions simultaneously and immediately determine an optimal D-plan and competitive D-plans. Direct dynamic methods are reviewed in Chaps. 5 and 6. The main drawback of the direct methods is unacceptably long calculation time and risk to lose the optimal D-plan, if alternative D-actions number n is large.

The essence of *iteration methods* is that dynamic tasks with many alternative D-actions solving is divided into phases—iterations. Each iteration has its own number of alternative D-actions and own optimization accuracy. Therefore, calculation of dynamic task with n alternative D-actions is replaced with calculation of v task with a smaller alternative D-actions number nv, which may result in considerable time economy for calculation of τ_α. For iteration methods realization, engineers' skill and erudition as well as formalized methods can be employed.

When applying iteration method, calculation time τ_α is decreasing. Calculation time, when solving n alternative D-actions by the direct method, can be expressed as follows:

$$\tau_a = k\mathrm{TM}q2^n, \tag{7.7}$$

where k coefficient, dependent on optimization method; TM number of optimization steps; q number of object nodes to be optimized.

In case of iteration method, calculation time may be estimated by the following formula:

$$\tau_a = vk\text{TM}q_v 2^{nv},\tag{7.8}$$

where q_v number of nodes in average iteration; nv alternative D-actions number in average iteration.

It is evident that optimization process partitioning into iterations gives more considerable calculation time savings than object (q) or D-steps number TM reducing.

In general case, in each iteration, the following operations are to be performed:

1. attribute optimization area;
2. work out a model of the object and D-process;
3. perform optimization;
4. analyze optimization results;
5. make a decision on optimization interruption (continuation).

Iteration process can be interrupted if the objective function fails to improve considerably. If the process must be continued, then alternative D-actions shall be structured into three groups:

1. D-actions, whose realization may be considered as absolute (in further iteration steps these are reviewed as attributed);
2. D-actions, which are inter-competitive, but it is not clear whether they are included in optimal D-plan;
3. D-actions, which are of no importance in optimization process (such actions in further iteration can be excluded from optimization area).

Formation of optimization area is selection of alternative D-actions. In the first iteration steps, only main D-actions may be considered, in further ones—less significant, second stage D-actions optimization area. If it is possible, relatively independent items can also be reviewed separately in the limited territory in each iteration.

The essence and possibilities of iteration method can be illustrated using functional and state effectiveness $B(t, e)$ of universal characteristic curve (see Fig. 7.5).

Let us assume that n are alternative D-actions with effectiveness b, but in optimization latest step TM, optimal state effectiveness opt $B(\text{TM})$ is reached at opt $m(\text{TM}) = (\text{opt}\,B\,(\text{TM}))/b$ realized D-actions. Then OIS number is calculated by formula

$$V_\omega = \sum_{m=0}^{\text{opt}\,m} C_n^m,\tag{7.9}$$

where C_n^m number of combinations of n elements taken m at a time.

Fig. 7.5 Optimal D-plan searching iteration processes. *1* functional *f*(*t*, *e*) deviation depending on state effectiveness B in optimization process with *n* = 30, opt *m*(*t*) = 5; *2* functional deviation in the first iteration with *n′* = 10, opt *m′*(*t*) = 2; *3* realized D-actions optimal sequence with letters marked states

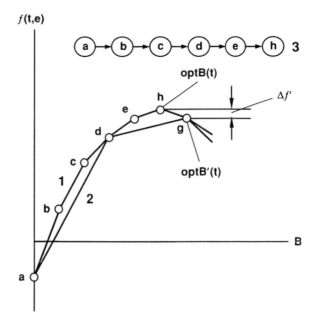

If $n = 30$ and opt $m = 5$, then $V_{\omega} = 170,000$. Exactly such task solution is not possible. Let us demonstrate how this task can be solved in two iterations.

The first iteration creates integrated D-actions—one integrated D-action consists of three D-actions, its effectiveness $b' = 3b$. The task becomes considerably less: $n' = 10$, opt $m'(\text{TM}) = 2$, but $V_{\omega} = 56$. Such optimization task can be solved directly. In optimization process, functional is changing along the broken line adg. In the first iteration, optimal solution (point h) is not obtained. The obtained solution (point g) functional differs from the maximal one by Δf.

In the second iteration, D-actions b, c, and d, which were selected in the first iteration, are excluded from the set of alternative D-actions. Then $n' = 27$, opt $m''(\text{TM}) = 2$, but $V_{\omega} < 500$. This task solution is real. In optimization process, functional is changing by the broken line adeh. As a result, real optimum is obtained.

There are certain disadvantages of the discussed iteration method. The formation of alternative D-actions in each iteration might not be formalized. Besides, the method is not operative as among iterations preparatory work for further iteration must be performed. Therefore, formalized iteration methods are required.

7.3.2 Dynamic Optimization Tasks Formalized Dispersing Method

The basis of the method is actions classification possibility with regard to system effectiveness. The method core is OIS method realization at limited OIS set, utilizing functional maximal effect function limit value. The principles of the

Fig. 7.6 OIS searching maximal effect method algorithm in case of limited OIS set, utilizing the critical value of functional maximal effect function (*blocks* are illustrated in Table 7.2)

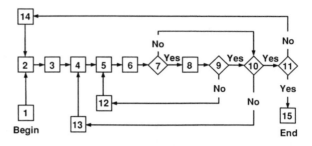

Table 7.2 The content of OIS searching block (Fig. 7.6) sub-blocks content in iteration and direct method

Block no.	Block content	
	Iteration method	Direct method
1	$t := 1$	
2	Step data and changeable set formation: $\Pi(t, v)$| opt $e(t, v-1)$	Step data formation
3	$\Omega(t) := 0, m := 0$	$\Omega(t) := 0, m := 0$
4	$\alpha := 1$	$\alpha := 1$
5	$e(m)_\alpha = $ opt $e(t, v-1) \cup d(m,v)$ formation	$e(m)_\alpha$ formation
6	$f(t,e(m))$ calculation	$f(t,e(m))$ calculation
7	max $Ef(t, e(m)) < \delta(v)$?	max $Ef(t, e(m)) < \delta(v)$?
8	$e(m)_\alpha$ incorporation $\Omega(t)$	$e(m)_\alpha$ incorporation $\Omega(t)$
9	$\alpha = C^m_{n(t,v)}$?	$\alpha = C^m_n$?
10	Is searching completed?	Is searching completed?
11	$t = T$?	$t = T$?
12	$\alpha := \alpha + 1$	$\alpha := \alpha + 1$
13	$m := m + 1$	$m := m + 1$
14	$t := t + 1$	$t := t + 1$
15	Is/are determined $\begin{aligned} & e(0) \subseteq \text{opt}\, e(1v) \subseteq \\ & \subseteq \ldots \subseteq \text{opt}\, e(t, v) \subseteq \\ & \subseteq \ldots \subseteq \text{opt}\, e(t, v) \end{aligned}$	Is/are determined $\begin{aligned} & e(0) \subseteq \text{opt}\, e(1v) \subseteq \\ & \subseteq \ldots \subseteq \text{opt}\, e(t, v) \subseteq \\ & \subseteq \ldots \subseteq \text{opt}\, e(t, v) \end{aligned}$

method are reviewed in Sect. 6.4; however, OIS set searching method of iteration methods must be improved.

First, let us review algorithm functioning at considerable variable number in case of direct optimization method application in steps $t = 1, 2,\ldots$, TM (see Fig. 7.6; Table 7.2).

In step t, the task of performed operations is to form OIS set Ω_t and retrace information. In the beginning, this set is empty (third block). OIS searching occurs in ascending sequence of realized actions number $m = 1, 2,\ldots$, opt m; besides, the number of formed states m corresponds to the number of combinations of n elements taken m at a time. Searching is finished if at m no optimal initial state appears. In such a way, formed states are marked with $e(t,m)_\alpha$. State functional is calculated with the help of recursive equation (Eq. 6.2).

The curve of functional $f(t, e)$ changes from m is $\cap$-shaped form, therefore, $e(t, m)_\alpha$ membership in Ω_t can be determined by the maximal effect function (Eqs. 6.26, 6.27) value.

If max $Ef(t, e(t, m)) < 0$, then state $e(t, M)$ is the optimal initial state. If we are working with criterion max $Ef(t, e(t, m))$, then all OIS are determined:

$$\max Ef(t, e(t, m)) < \delta. \tag{7.10}$$

In such case not all OIS are determined, but only those which are formed realizing the most effective actions.

In order to obligatory determine an optimal plan in all D-steps, all OIS must be stored. Yet, due to their considerable number, it is not possible. In practical tasks, it is required to decline all OIS which got smaller functional values.

The optimal plan is lost if at least in any step its state is not preserved. A possibility of losing the optimal plan is increasing in tasks, where several main strategies are considered with a great variation number, created by relatively tiny actions. Variations by functional value differ slightly. These differences are marked with Δf_v. Differences with regard to functional major strategies are marked with Δf_S that is much greater than variations difference: $\Delta f_S \gg \Delta f_v$. If Δf_S is approaching Δf_ω (functional changes within the range of OIS set Ω), then a competitive strategy has been lost in intermediate steps. In such case direct methods results are close to those of quasi-dynamic optimization.

In Fig. 7.6 an algorithm is presented, which functions at maximal effect function marginal values $\delta < 0$; its characteristic feature is that Δf_ω increases, but the number of states with various distinctive actions reduces sharply. As a result, OIS set Ω comprises major strategies. The mentioned characteristics create pre-conditions for formalized iteration methods elaboration.

Iteration method algorithm is shown in Fig. 7.7. Iteration step core is multi-step $t = 1, 2,\ldots$, TM optimization with limited OIS set.

Iteration step v is characterized with these step parameters:

$\Pi(t,v)$ Alternative actions set that changes by D-steps and iteration steps;
$n(t,v)$ Alternative actions number;
$\delta(v)$ Maximal effect function critical value in iteration step v;
$\Delta\delta$ Maximal effect function critical value changeable step.

In iteration step, dynamic optimization results in the optimal plan which may be expressed as related states sequence:

$$e(0) \subseteq \operatorname{opt} e(1, v) \subseteq \ldots \subseteq \operatorname{opt} e(t, v) \subseteq \ldots \subseteq \operatorname{opt} e(\mathrm{TM}, v). \tag{7.11}$$

State opt $e(\mathrm{TM},v)$ corresponds to functional maximal value $\left(\max_{e \subseteq \Omega(\mathrm{TM})} f(\mathrm{TM}, e, v) \right)$.

Fig. 7.7 The flowchart of the algorithm for dynamic optimization OIS iteration method

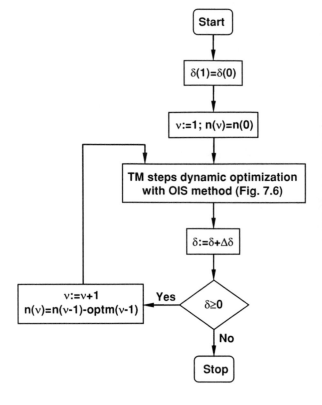

The states that form an optimal plan in iteration $v - 1$ are used to obtain next iteration step parameters. Step parameters in iteration process change according to these recursive equations:

$$\Pi(t, v) = \Pi(t, v - 1)\backslash \text{opt}\, e(t, v - 1), \qquad (7.12)$$

$$e(v)_\alpha = \text{opt}\, e(t, v - 1) \cup d(m, v)_\alpha, \qquad (7.13)$$

$$\delta(v) = \delta(v - 1) + \Delta\delta, \qquad (7.14)$$

where $d(m,v)_\alpha$ combination of m realized actions from $\Pi(m,v)$.

At the beginning of dynamic optimization task solving, set $\Pi(0)$ and values $n(0)$, $\delta(0)$ and $\Delta\delta$ are known. Figure 7.8 illustrates the realization principle of optimization with iteration method and the method effectiveness.

Two iteration tasks and optimization process characteristic curves are depicted in the illustration. In the first iteration, maximal effect function critical value is $\delta(1)$; besides, searching depth is decreasing $m_{\text{sear}}(t,v) < \text{opt}\, m(t)$, also calculation time and required memory is reducing: $\tau_{\alpha 1} < \tau_\alpha$ and $V_{\omega 1} < V_\omega$. In the second iteration, variables number is decreasing: $n(2) = n(0) - m_{\text{sear}}(1)$, but optimization accuracy $\delta(2) = 0$, OIS searching depth $m_{\text{sear}}(2)$, calculation time $\tau_{\alpha 2}$ and $V_{\omega 2}$ are

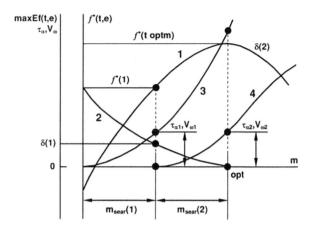

Fig. 7.8 Characteristic curves for dynamic optimization iteration method. *1* relative functional value $f^*(t,e)$ dependence on realized actions number; *2* maximal effect function value max $Ef(t,e)$ dependence on m; *3* estimation time τ_α and utilized memory V_ω dependence on searching depth in case of n variables; *4* estimation time τ_α and utilized memory V_ω dependence on searching depth in the second iteration with $n(2) = n - m(1)$ alternative actions; $m_{sear}(1)$ and $m_{sear}(2)$ searching depth for the first and second iteration

increasing. Searching ends (see Fig. 6.5) if condition max $Ef(t, e\ (t,m)) > 0$ is fulfilled, i.e., at searching depth opt $m = m_\alpha(1) + m_\alpha(2)$. The total estimation time is $\tau_{\alpha 1} + \tau_{\alpha 2} < \tau_\alpha$, where τ_α estimation time utilizing the direct method.

Iteration method effectiveness is determined by the circumstance that in each iteration searching depth of OIS is decreasing. Concurrently, actions classification with regard to their system effectiveness is taking place—in the first iteration there are selected D-actions with higher system effectiveness but in the next—actions with less system effectiveness, which improves previously selected states. Iteration methods for tasks with a large number of actions outperform direct optimization methods. It becomes apparent also in more adequate optimal plans. Besides, system development management operative capability is improved as calculation operations are performed quicker.

7.4 Optimization Method in Adaptation Period

Optimization period TM is subdivided into two groups: planning period first t_p steps, for which decisions are being, and residual period beyond it up to estimation period expiry in adaptation period $t = 1, 2,...,t_p, t_p + 1,...,$ TM. Key persons in charge for planning are often interested to figure out only D-plan for planning period. Still economic and technical effectiveness must be estimated in each D-plan for the entire estimation time period. Comparing definite alternative D-plans for the planning period, network development in further steps (beyond planning period up to the end of estimation period) must be estimated as well. Dynamic adaptation is a method, providing supplementary actions, which are to be realized in adaptation period in order to ensure D-plan optimal continuation up to the end of estimation period. According to planning period $t = 1, 2,..., t_p$ and adaptation period $t = t_p + 1,...,$ TM, various recursive equations must be utilized. For instance, in planning period, recursive equation (Eq. 5.4) must be used, while in adaptation period $t = t_p + 1,...,$ TM distinctive recursive equation is applied:

$$f(t, e(t)) = f(t - 1, e(t - 1)) + \max_{\{e(t-1) \subseteq e(t)\}} F(t, e(t)). \qquad (7.15)$$

The purpose of optimization in adaptation period is to find an optimal continuation for each D-plan determined (or given) in planning period by recursive equation (Eq. 7.14).

Application of various optimization methods for planning and adaptation period is motivated by convincing evidence. Using Eq. 5.4 for the entire estimation period can lead to losing D-plans determined in the planning period. Adaptation method can be utilized, revising user assigned D-plans too. In order to present dynamic adaptation method, let us consider a simple example.

A 20-kV distribution network (see Fig. 7.9) is located within the distance of 30 km from 110/20 kV power supply substation. New consumer must be connected to the network—sawmill with initial load 3 MW, which has doubled within a 20-year period. Thus, the task is to determinate electric power supply scheme for the new consumer considering a 25-year estimation period.

The estimation period is to be divided into four steps. Power flow analysis demonstrates that there are inadmissible voltage losses in the first step, but in the third step the transformer is overloaded.

Alternative D-actions for electric power supply reinforcing are presented in Fig. 7.9 and Table 7.3.

Planning period is the first two D-steps. In the first step, there are three potential states with the following functional values:

$$f(1, \{1\}) = 2,664, \quad f(1, \{2\}) = 3,173, \quad f(1, \{1,2\}) = 2,767.$$

In correspondence with OIS method, only D-states {1} and {2} are OIS which are involved in optimization process. In the second optimization step we obtain

$$f(2, \{1\}) = 2,970, \quad f(2, \{2\}) = 3,173, \quad f(2, \{1,2\}) = 3,046.$$

Fig. 7.9 Alternative D-actions providing electric power supply for new consumers

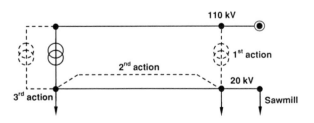

Table 7.3 Alternative D-actions

No.	Type of D-action	Expenditures m.u. $\times 10^3$
1	Construction of new substation	400
2	Construction of new line	153
3	Reinforcement of existing substation - transformer replacement in the forth step	40

In the second optimization step, OIS are {1} and {2}. Thus, in planning time period, two alternative D-plans are obtained:

1. first D-action realization (construction of new substation),
2. second D-action realization (construction of new line).

In both D-plans, the D-actions are realized in the first D-step. The D-plans can be represented by states sequence:

first D-plan: {0} ⊂ {1} ⊂ {1},
second D-plan: {0} ⊂ {2} ⊂ {2}.

Network D-process is illustrated graphically in Fig. 7.10 where t is D-step, {1,3} denotes D-state with realized D-actions 1 and 3, and arrow denotes transition from one D-state to another. Non-interrupted arrow line indicates the D-plans obtained in optimization process.

Starting with the third step, each D-plan, which was obtained in planning period, is then supplemented so that functional least possible augmentation is ensured. The first D-plan follow-on in the third and forth steps, looks like this:

$$f(3, \{1\}) = 3,424, \quad f(3, \{1, 2\}) = 3,526;$$

Fig. 7.10 Network development graph

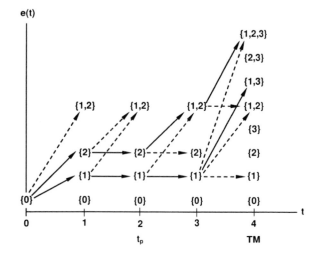

$$f(4, \{1\}) = 1,144 \times 10^3, \quad f(4, \{1,2\}) = 1,175 \times 10^3;$$

$$f(4, \{1,3\}) = 3,536, \quad f(4, \{1,2,3\}) = 3,558.$$

Thus, optimal follow-on of the first D-plan in the last step is D-state $\{1,3\}$. The first D-plan for complete estimation period can be expressed as the following D-states sequence:

$$\{0\} \subset \{1\} = \{1\} = \{1\} \subset \{1,3\}.$$

In similar way, continuation of the second D-plan is obtained:

$$f(3, \{2\}) = 4,121 \times 10^3, \quad f(3, \{1,2\}) = 3,728;$$

$$f(4, \{1,2\}) = 1,174 \times 10^3, \quad f(4, \{1,2,3\}) = 3,662.$$

It is followed by the second D-plan, which is expressed in the form of these D-states sequence:

$$\{0\} \subset \{2\} = \{2\} \subset \{1,2\}.$$

In the last (forth) step, the first D-plan is more expedient than the second. To make a final decision under uncertain information conditions, risk analysis must be performed.

Chapter 8
Information Technology for Sustainable Development of Electric Power Systems

Abstract This chapter presents basic principles of information technology for sustainable development management of Latvian power system. The condition and development tasks of Latvian power system are reviewed. In this chapter, we also offer basic principles of LTS development optimization. The structure of information technology, as well as main module functions and brief characteristics of informative links amongst modules are presented. In this chapter, we analyze functioning module of electrical power plant and functioning module of transmission network. The main parts of functioning module of electrical power plant are (a) energy source model, (b) source operation method and software. The main parts of functioning module of transmission network include: (a) network model for power flow calculation and (b) consumer load (demand) model.

8.1 Information Technology Functions and Tasks to be Solved

Chapters 8, 9 and 10 consider information technology (IT) of sustainable development management of electric power systems transmission network and generation sources *LDM-TG'08*. It serves as an example of real system sustainable development management that is based on OIS method and maximal effect OIS searching method [1, 2, 8, 9]. The reviewed example can also be used by the readers who are developing sustainable development management technologies for other systems. In order to develop information technology for a particular system, the appropriate model of this system functioning at D-step level (see this Chapter) must be developed, as well as data and knowledge base (see Chap. 9), and particular system operation program (see Chap. 10).

Elaborating the mentioned information technology, the key target was to create new transmission network and generation system development management concept and methods to validate the made decisions when simulating situations

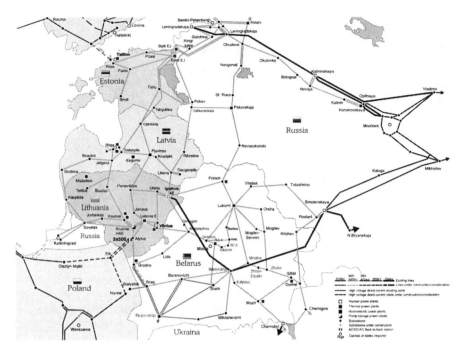

Fig. 8.1 Operation of Baltic power systems in electric power ring

and their respective probable consequences thus forwarding background for optimal D-plan selection. In order to ensure transmission network and generation sustainability, it is necessary to precisely analyze consequences of the made decision while developing project design of electric power system. Therefore, in design of this information technology dynamic methods are utilized, enabling one to observe D-process and far perspective, if required, for the period up to 25 D-steps (50–70 years).

Solving the task on the Latvian development management of electric power system, all integrated power systems of the Baltic states must be analyzed, taking into consideration the existing and envisaged in perspective external links.

Latvian power system is operating in power integrated union of the Baltic power systems—of Estonia, Latvia and Lithuania—330 kV electric power networks with 330 kV transmission power network and concurrently these form some of the electric power ring components incorporating 330, 500 and 750 kV power networks that belong to neighboring states power systems—either Belarus or Russia (see Fig. 8.1).

330 and 750 kV electric power ring significantly improves power supply reliability of each state power system, if its development is coordinated in the frame of large integrated power system. There are electric power links with neighboring power systems, the so-called electric power cross-sections, which are limited by

maximal admissible transmission line capacity in normal and post-emergency operational states. The problem of electric power ring is not to be solved in local separate power systems as thee solutions are neither effective, nor sufficiently economically motivated, and fail to guarantee reliable and secure power supply to consumers. The problems of electric power ring can be solved effectively only by the joint efforts of all the three Baltic states. Affiliated to the European Union, the power system of Latvia is open to liberalized electricity market and opportunity to connect their networks to the grid of the European countries. As the neighboring European countries close to Latvia are Lithuania and Estonia and there is no other output access to the European grid, the power system of Latvia participates in the joint projects of the Baltic states on construction of new lines with Western European states. There are already two such projects: underwater direct current cable line between Estonia and Finland across The Baltic Sea and two 400 kV transmission lines between Lithuania and Poland (Alitus-Elk). This connection is required to reinforce power supply reliability of the three states and extend liberalized electricity market, where each state is envisaged to be a competitive participant, because liberalized electricity market would be attractive and profitable either for Lithuania, Latvia or Estonia to acquire more or less considerable possibilities to sell surplus of electricity and electricity quality improving service opportunities.

Information technology is focused on the improvement of investment effectiveness and action substantiations on prime costs reduction in high-voltage networks. Information technology can also be applied for development planning in network utilities, project design organizations, educational and scientific research institutions. Information technology performs economic analysis and optimization observing development perspective (20–50 years) and initial information uncertainty.

With such tool as information technology, it is possible to:

1. Estimate technical economic state of the given system including identification of bottle-neck problems;
2. Determine whether the system is economically effective to

   - Construct new power plants (with what capacity and where?);
   - Reconstruct or upgrade the existing power plants;
   - Construct new substations and electric transmission power lines;
   - Dismantle or reconstruct substations and electric transmission lines;
   - Transfer to higher nominal voltage in the given network;

3. Determine the priority of various actions in case of limited investment resources.

Information technology enables one to analyze a network with several nominal voltage levels (up to five), performing both technical economic analysis and optimization of it.

Information technology for each D-plan calculates the *annual technical economic criteria*:

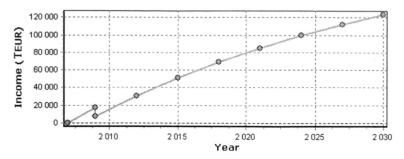

Fig. 8.2 Investment pay-back time Schedule

1. Electric energy consumption per year;
2. Power balance for all operation states;
3. Investment and residual investment value at the end of period;
4. Operation costs:

   – Maintenance operation costs;
   – Repair expenditures;

5. Losses costs;
6. Energy non-delivery costs;
7. Energy costs in power plants;
8. Income resulting from electric energy sales;
9. Losses in network:

   – Power losses;
   – Energy losses;
   – Energy losses in percentage;

10. Penalty function technical validity; and total technical economic criteria:

    – Objective function total reduced value;
    – Total reduced investments;
    – Total residual value;
    – Total reduced operation costs;
    – Total reduced losses costs;
    – Total reduced power plants costs;
    – Total reduced non-delivery costs;
    – Total reduced income;
    – Technical validity.

The investment pay-back time schedule can be reviewed for any D-plan (see Fig. 8.2 for illustration).

Information technology maximal parameters are the following:

Optimization steps number	25
Number of subscriber groups	20
Number of OIS	200
D-plans number	10
Links number	1,000
Nodes number	600
Number of circuit breakers parameters	100
Switchgears number	150
Data array number for network reliability estimation	300
Failures (faults) number—sections number	500
Number of operational states	15
Number of links utilized in D-actions, new (+) and liquidated (−)	200
Number of closed objects in table	600
Number of concurrent outages	100
Number of voltages	5
Number of power plants	100
Maximal average points number of compound links	20
Number of switchgear components	40
Number of alternative D-actions	30
Maximal number of OIS	250

8.2   Basic Principles of System Optimization

The principal theses for the concept of electric network development optimization are elaborated in line with liberalized electricity market key issues, functions and elements as well as with analysis results of network sustainability aspects. The first and most essential conclusion obtained as a result of analysis is that the significance of network planning and optimization is steadily increasing under conditions of liberalized electricity market. However, the existing methods must be revised and supplemented to a great extent; only then beneficial results of development planning can be gained. We have elaborated seven principal theses, which summarize our long experience in network development planning, and we recommend using them in elaboration process of new development optimization methods. Let us review these principal theses in more detail.

1. *Optimization must be dynamic*—regular and flexible (adjustable).

Development of real electric network is traditionally non-interruptible process that cannot be replaced by 1–3 development horizons. As our research demonstrates, a model like that distorts plans of technical economic comparison. Under conditions of liberalized electricity market, only the model of multi-step D-process must be utilized. According to the method, economic analysis must be done in regard to the object economic life-cycle (in electric network it is 20–50 years).

Fig. 8.3 User-guided quality structure

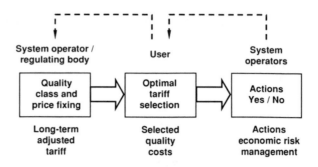

Objective function must be calculated for the entire estimation period T. Under information uncertainty conditions, the final decision must be made only for acceptable short perspective (3–5 years). Let us refer to this time period as *decision making period* $t_d < T$.

Since information on development in estimation period is uncertain, the decisions made are to be regularly updated. The designing process of development has to be gradual and contain certain phases such as:

- External information updating;
- Utilizing updated information on previous system development.

2. *In optimization, it is required to take into consideration system sustainability, examining the far perspective with a view to estimate the possible consequences of the decision made.*

In case of disregarding of electric network sustainability aspects, estimation period is 20–30 years, which sufficiently accurately can be modelled with 10 D-steps; but taking sustainability aspects into consideration, it is required to review the period of 50 years or 20 D-steps.

3. *In optimization, power supply reliability and security must be taken into consideration to stipulate requirements to independent electricity producers.*

The main problem of reliability estimation is that the prescribed quality level of electric power supply requires certain costs maintained by owner, while quality fault or shortage influence costs of consumers. Electric power supply quality— along with price—is the most interesting and tricky description of electric energy, particularly in regard to industrial consumers.

From the consumers' point of view, it would be preferable to establish a structure where customer can select individual reliability level and associated costs. Figure 8.3 illustrates general principles of the approach when customers select reliability level paying for connection and service as well as insurance for additional reliability.

Reliability estimation methods are subdivided into two groups: short-term and long-term.

The first group is utilized for power system and networks on-line operation management. The second is envisaged for development planning including networks development optimization.

Commendatory reliability criteria on networks development optimization are as follows:

1. Non-delivered energy;
2. Costs of non-delivery;
3. Power supply outage time.

Reliability criteria must be calculated for line (considering all lines disconnections in succession and assuming that overloaded lines must be disconnected) and for substations and switchgear units.

4. *Information uncertainty must be taken into consideration in optimization, namely*:

   1. Physical uncertainties,
   2. Financial uncertainties,
   3. Regulator uncertainties.

Selecting D-plans under uncertainty conditions occurs as follows: (1) select information package set which represents information credibility range; further on package *i* is referred to as forecast l; (2) select D-plan estimation criteria; (3) select D-plans for comparison; (4) select optimal D-plan.

When estimating D-plans under uncertainty conditions, based on numerous researches, it is sufficient to have a comparatively small information package that represents information credibility range. This is due to the following aspects: objective function curve in optimum area is very flat and system parameters change discretely.

For electric networks development optimization tasks, the most applicable criterion is the minimal maximal risk

$$\min_{j \in v} R(j)_{\max} = \min_{j \in v} \max_{i \in \mu} (F(j, i) - F(i)), \tag{8.1}$$

where F(*i*) is the maximum objective function value in case of forecast *i*.

Adaptation Decision making and estimation period differentiation ($t_d < T$) demands radically that D-plan definition has to be altered—D-plan is characterized only by D-actions realized in decision making period.

D-actions, which are realized beyond decision making period, are obtained in risk analysis in the course of adaptation (dynamic optimization).

5. *In optimization, networks and generation must be considered concurrently*

In order to model energy source, network model must be supplemented with generator nodes and power plant links (see Sect. 8.4.1).

With the help of power plant links, energy saving technologies can be simulated too, as well as liberalized electricity market and electricity import from neighboring power systems. Solving development tasks of networks and generation systems concurrently, annual load and generation schedule is modelled with coordinated operational states. It is advisable to consider 13 operational states—each month corresponds to one operational state: (1) January, (2) February, ..., (12) December, (13) post-emergency. Such model allows one to review systems with hydropower plant (HPP), wind power plant (WPP) and cogeneration heat plants (CHP) whose output is influenced by meteorological conditions.

Concurrent network and generation system dependence in optimization process on D-plan, D-step and operational state changes the number of operable energy sources and structure. In order to be able to compare various D-plans objectively, these must be examined at relevant optimal load distribution amongst energy sources. Such optimization calculation must be performed for all D-plans in all D-steps and all operational states.

6. *Specialized information technologies must be applied in optimization.*

Information technology utilization provides opportunity to:

- Increase planning effectiveness, standardizing its processes and applied software, and automating work with software;
- Reduce information technology system maintenance costs, replacing many programs by some high integrated and advanced software packages.

8.3 Information Technology Structure and Main Module Functions

Main modules structure is represented in Fig. 8.4.

Model for external data attachment to information technology database. Information technology also utilizes external database that allows one to quickly and precisely form existing state of electric power system. As external database, transmission network operational state calculation database is utilized.

Module *Actions* forms development model. Development model is utilized to determine optimal D-plan by estimation method (*Plans*), as well as by optimal initial state method (*optimization*).

The important notion when applying network economic analysis complex is possible (alternative) D-actions construction, reconstruction or dismantling of power plants, lines or substations. "Actions" may be interpreted as "purchases". Actions are always related to activities. Such variables differ from the traditional approach in many other optimization modules where parameter elements are modelled. We are searching for "what to do" and "when to do" parameter elements which are created as a result of activities [3–7].

Module *Network state estimation*: (1) load flow calculation, (2) switchgear reliability estimation, (3) network reliability estimation, (4) energy balance calculation for each state.

Fig. 8.4 Main modules
structure

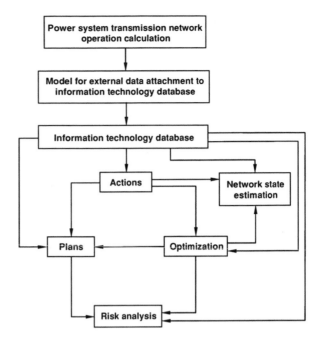

Module *Plans* can be utilized for optimal D-plan estimation, when comparing various D-plans as well as for reduction of the number of D-actions, which are utilized by user in optimization module.

Module *Optimization* determines competitive D-plans set (up to ten D-plans). In Optimization module, OIS method is utilized. Optimization is performed under deterministic information—average forecast.

Module *Risk analysis* is utilized for optimal D-plan selection under uncertain information conditions.

8.4 Subsystem Functioning Models

8.4.1 Functioning Model of Electric Power Plants

Energy source model. In order to simulate energy source, new generator node is to be added to system model. Generator node and network node, to which energy source is added, are connected to electric power plant link (see Fig. 8.5).

Energy saving technologies and electric energy import of neighboring power systems can be modelled with the help of electric power plant link. Solving tasks of energy source and network development concurrently, annual loads and generation schedules are modelled with several operational states, for example, winter maximum, winter night maximum, winter night minimum, and operational state in power system with HPP when HPP operation is terminated.

Fig. 8.5 Energy source
model

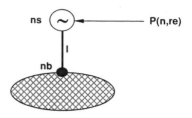

Information technology in each D-step may review up to 15 operational states. The states are characterized by:

- Capacity in load nodes,
- Admissible generation in generator nodes, $P(n,\text{re})$,
- Operational state total duration in hours/year, $\text{Tr}(\text{re})$.

Electric power plant link l is to be designated with:

- Blocks number $sk(l)$,
- Block nominal capacity $Pn(l)$, MW,
- Specific fuel consumption for electricity production $b(l)$, $\text{kWh}(t)/\text{kWh}(e)$,
- Relative fuel transport costs $\text{transp}(l)$, r.u.,
- Fuel costs $cf(l,t)$, m.u./$\text{kWh}(t)$,
- Outage duration for repair $\text{FOR}(l)$.

If instead of fuel costs, energy selling price is fixed as EUR/10 kWh (e), then $b(l) = 1$, $\text{transp}(l) = 1$.

Sources operational states optimization—statement and method Energy sources number and structure are changing depending on D-plan and D-step in concurrent energy source and network optimization process. In order to be able to compare various D-plans, these must be considered for relevant optimal load distribution amongst energy sources. Information technology performs such optimization calculation for all D-plans, in all steps and all operational states. Optimal generation $PL(l,\text{re})$ is determined for electric power plants links so that total expenditures for fuel in all operational states would be minimal

$$\text{Cfl}(t) = \sum_{\text{re}=1}^{\text{rem }(t)} \sum_{l \in \text{MG}} \text{Czfl}(t,l) \rightarrow \min, \qquad (8.2)$$

where fuel costs for source l in operational state re are:

$$\text{Czfl}(r,l) = Pl(l,\text{re}) \cdot \text{Tr}(\text{re}) \cdot b(l) \cdot \text{transp}(l) \cdot \text{Cf}(l,t). \qquad (8.3)$$

Optimization observes these limitations:

1. $Pl(l) \le P(n,\text{re}) \cdot sk(l)$ $l \in \text{MG}$,
2. $\Sigma Pl(l) = \text{Psd}(\text{re})$,

where $\text{Psd}(\text{re})$ is power system total demand (load) in operational state re.

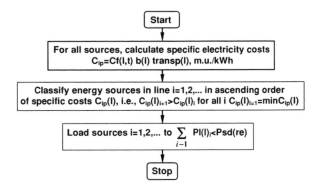

Fig. 8.6 The flowchart of sources operational state optimization algorithm

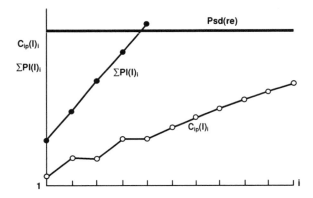

Fig. 8.7 Optimization scheme of sources operational states

The flowchart of sources operational state optimization algorithm is shown in Fig. 8.6, and optimization scheme in Fig 8.7.

Performing flow calculation with altered costs cf or/and P_{max} for one or several electric power plants—the results of Block 2 and Block 3, i.e., sources load, may vary.

Loads and output mathematical model It is evident that estimation operational states depend on various factors: e.g., hydropower plant, on water flow and water reservoir volume; wind power plant, on wind speed and number of windy days per month; and cogeneration heat plants, on ambient temperature in winter. In order to obtain flow calculation results that illustrate real situation, mathematical model of special loads and electric power plants output is required. Loads and outputs must be correlated and coordinated and these must reflect meteorological conditions (methodology of load schedules formation is presented in Sect. 8.4.2).

Methodology for electric power plant output schedule formation In order to obtain coordinated transmission network capacity calculation, the power plant generation output schedules must be created in the same way as demand load schedules. Let us review 12 operational states with monthly average outputs expressed by related units of maximal capacity. Average output is obtained by the measurement result during long-term period. Time period for all operational states

Table 8.1 Data on average monthly output by electric power plants, MW

EPP type	Nom. capacity	Average relative output by month											
		1	2	3	4	5	6	7	8	9	10	11	12
CHP	400	1.0	1.0	1.0	0.0	1.0	1.0	1.0	1.0	1.0	1.0	1.0	1.0
NPP	300	1.0	1.0	1.0	0.0	1.0	1.0	1.0	1.0	1.0	1.0	1.0	1.0
CHP	600	1.0	1.0	1.0	1.0	1.0	1.0	1.0	1.0	1.0	1.0	1.0	1.0
CHP	800	1.0	1.0	1.0	1.0	1.0	1.0	1.0	1.0	1.0	1.0	1.0	1.0
CHP2	390	0.7	0.5	0.4	0.1	0.1	0.2	0.2	0.2	0.2	0.3	0.3	0.3
HPP1	262	0.2	0.3	0.5	0.7	0.3	0.2	0.1	0.1	0.1	0.1	0.2	0.4
HPP2	847	0.2	0.2	0.5	0.8	0.2	0.1	0.1	0.1	0.1	0.1	0.2	0.3
HPP3	402	0.1	0.2	0.4	0.7	0.2	0.1	0.1	0.1	0.1	0.1	0.2	0.3
CHP1	129	0.5	0.4	0.3	0.2	0.1	0.0	0.0	0.0	0.0	0.1	0.3	0.4

Fig. 8.8 Electric power plants annual output schedule

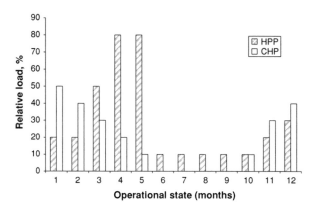

is 730 h/year. Data about output of electric power plants in the Baltic states are summarized in Table 8.1.

Figure 8.8 shows data on output annual schedules for these electric power plants: hydro power plant (HPP) and cogeneration heat plant (CHP).

The load schedules illustrated evidently demonstrate that power plants (both HPP and CHP) output distribution within a 1-year period is very irregular (unbalanced).

In the analysis of a particular object, real data on output schedules must be utilized.

Network calculation operational states Calculation operational states within a 1-year period in Latvian power system transmission network are quite diverse. The operational states analyzed in the example are given in Table 8.2.

It is obvious that loads of electric power plants are related to calculation operational states. Possible number of calculation operational states is 15. Output optimization is performed automatically according to expenditures. The maximal potential electric power plant capacity in operational state is determined by energy source (wind, water, heat) resources.

Table 8.2 Calculation operational states

No. of operational state	Name	Tre, hours/year	$\sum P$	$\sum P_g$	$\sum P_{imp}$	Overload 110 kV	Overload 330 kV
1	January	730	1,192	553	639	9	0
2	February	730	1,081	576	505	10	0
3	March	730	1,192	866	325	12	0
4	April	730	934	934	0	12	0
5	May	730	995	392	603	14	1
6	June	730	872	267	604	7	1
7	July	730	946	206	739	11	1
8	August	730	934	185	748	11	1
9	September	730	1,020	215	805	13	1
10	October	730	1,118	357	761	13	1
11	November	730	1,179	558	621	13	1
12	December	730	1,229	645	583	12	1

The suggested method can be utilized in research on power engineering systems with various types of power plants, as well as for investigating the impact of private electric power plants (disperse generation) on 110 kV network (lines loading, energy losses, etc.) [1].

8.4.2 Transmission Network Functioning Model

Transmission network functioning determines power flow in changeable scheme of network where as a result of D-actions realization, new power supply substations and lines appear. In such conditions, power flow optimization must be done too. Power flow optimization and power flow calculation in the frame of electric power system development dynamic optimization task must be performed with high speed and certain accuracy. This is the main criterion that determines the selection of functioning model.

Power flow and voltage loss calculation method for *multilevel* network is illustrated in Fig. 8.9.

Network model consists of subsystems. One subsystem comprises one voltage node and line links. Subsystems are interconnected by transformer links. Node potential calculation is performed gradually by subsystems starting with lower nominal voltage.

In the process of dynamic optimization in distribution network, voltage quality must be estimated taking the following into consideration: nodes number with inadmissible voltage levels, inadmissible voltage loss values, supplementary resources for voltage regulation, and loads, produced by non-adequate quality voltage value. Maximal voltage losses are division points. If division points are not selected optimally, the voltage losses might be much more considerable as

Fig. 8.9 Network model for power flow calculation

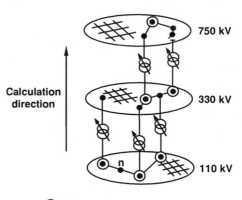

Calculation direction

750 kV

330 kV

110 kV

◉ - Feeding centre ● - Load node

compared to the case of optimal point selection. Thus, comparing D-plans with various substation number and locations, the optimal division points must be determined for each D-plan. This can be modelled in the following way. Each voltage level is considered as closed circuit network. The feeding substation busbars are assumed to be nodes with the fixed voltage level. In such model, nodes with varied power supply correspond to optimal division points in which there is maximal ΔU. Therefore, nodes in network model for power flow calculation can be classified into feeding centres, load nodes, and generation nodes.

Feeding centres are feeding nodes and lower voltage busbars existing in network transformers.

Node n potential δ_n is voltage losses from feeding centre up to the node in related units.

Active power flow in link Pl_{ij} is calculated by formula (positive flow direction from initial node i to end node j):

$$Pl_{ij} = Y_{ij} \cdot (\delta_i - \delta_j), \tag{8.4}$$

where Y_{ij} is the conductivity coefficient of link ij.

$$Y_{ij} = \frac{U^2}{R_{ij} + X_{ij} \cdot tg}. \tag{8.5}$$

Kirchhoff's first rule for node n can be expressed as:

$$\sum_{ij \in V_n} a_{ij} \cdot Pl_{ij} + P_n = 0, \tag{8.6}$$

where V_n—for node n incident links in total (list); a_{ij} —coefficient; $a_{ij} = 1$, if n is links ij initial node, otherwise $a_{ij} = -1$; P_n —load in node n, observing loads in transformer links.

Kirchhoff's second rule for circuit C may be expressed as:

$$\sum_{ij\in C} \Delta U_{ij} = 0. \tag{8.7}$$

Observing expressions 8.4–8.7, a system of linear equations for node potential δ_i calculation may be written as:

$$
\begin{aligned}
-Y_{11}\delta_1 + Y_{12}\delta_2 + &\cdots &+ Y_{1i}\delta_i + &\cdots &+ Y_{1j}\delta_j + &\cdots &+ Y_{1m}\delta_m = P_1 \\
Y_{21}\delta_1 - Y_{22}\delta_2 + &\cdots &+ Y_{2i}\delta_i + &\cdots &+ Y_{2j}\delta_j + &\cdots &+ Y_{2m}\delta_m = P_2 \\
&\cdots & \cdots \quad\quad & \cdots & \cdots \quad\quad & \cdots & \\
Y_{i1}\delta_1 + Y_{i2}\delta_2 + &\cdots &- Y_{ii}\delta_i + &\cdots &+ Y_{ij}\delta_j + &\cdots &+ Y_{im}\delta_m = P_i \\
&\cdots & \cdots \quad\quad & \cdots & \cdots \quad\quad & \cdots & \\
Y_{m1}\delta_1 + Y_{m2}\delta_2 + &\cdots &+ Y_{mi}\delta_i + &\cdots &+ Y_{mj}\delta_j + &\cdots &- Y_{mm}\delta_m = P_m
\end{aligned}
\tag{8.8}
$$

where Y_{ii}—node i self-conductivity coefficient

$$Y_{ii} = \sum_{ij\in V_i} Y_{ij}$$

V_i—incident links set for node i.

Consumer load (demand) schedule formation method The loads of transmission network are determined by 110 kV substation loads. These loads are regularly measured, thus, it is easy to create load schedule corresponding to monthly average load. With average load, power and energy losses in electric networks can be calculated sufficiently precisely. In Table 8.3 and Fig. 8.10, load schedules of 110 kV substations of Latvia are illustrated.

For estimation of network overloading, it is required to perform calculations at the maximal load. If average load P_v is known, the maximal load can be calculated by formula:

Table 8.3 Consumer load schedules (P1, P2, ..., P5 indicate consumer type according to annual load schedule)

No.	Name	Tre, hours/year	P1, r.u.	P2, r.u.	P3, r.u.	P4, r.u.	P5, r.u.
1	January	730	0.97	1	0.32	0.23	0
2	February	730	0.88	0.88	0.69	0.19	0
3	March	730	0.97	0.87	0.69	0.19	0
4	April	730	0.76	0.65	0.56	0.16	1
5	May	730	0.81	0.62	0.56	0.16	0
6	June	730	0.71	0.55	0.52	0.13	0
7	July	730	0.77	0.58	0.55	0.14	0
8	August	730	0.76	0.58	0.62	0.13	0
9	September	730	0.83	0.64	0.62	0.16	0
10	October	730	0.91	0.77	0.81	0.23	0
11	November	730	0.96	0.86	0.95	0.8	0
12	December	730	1	0.93	1	1	0

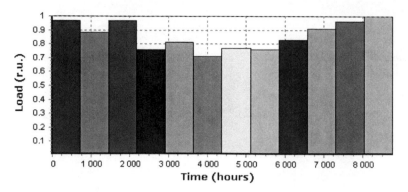

Fig. 8.10 Load annual schedule P2

$$P_{max} = \frac{P_v \cdot 8760}{T_{max}}, \qquad (8.9)$$

where T_{max} is the maximal load time.

In the integrated power system of the Baltic states, $T_{max} = 5,000–7,000$. It means that $P_{max} = P_v k$, where k is the load maximum coefficient that is varying within the interval from 1.3 to 1.7. Performing power flow calculation, it is possible to take this coefficient into account.

References

1. Krishans Z, Oleinikova I (2001) Software for power supply utilities management. Riga Technical University, Riga, Latvia (in Latvian)
2. Krishans Z, Oleinikova I, Mutule A (2004) New technologies for power system long-term management,Siktivkar, Russia. Управление электро-энергетическими системами - новые технологии и рынок pp 133–139
3. Krishans Z, Oleinikova I, Mutule A, Kutjuns A (2006) Optimization method of power system development under uncertainty. In: 9th international conference on probabilistic methods applied to power systems, (symposium proceedings on CD) KTH, Stockholm, Sweden
4. Krishans Z, Oleinikova I, Mutule A, Runcs J (2006) Method for power system modernization asset management optimization under liberalization. In: 3rd International workshop proceedings liberalization and modernization of power systems: Risk assessment and optimization for asset management, IEEE Power Engineering Society, Irkutsk, Russia, pp 36–41
5. Krishans Z, Oleinikova I, Mutule A, Runcs J (2006) Dynamic simulation method for transmission and distribution planning. In: 5th WSEAS International conference on system science and simulation in engineering (Conference Proceedings on CD), Tenerife, Canary Islands, Spain, pp 25–30
6. Krishans Z, Mutule A, Oleinikova I,(2006) Structure of optimisation models system for planning of power system under market conditions. In: 7th International scientific conference on "Electric Power Engineering (EPE) 2006," (Conference Proceedings) Brno University of Technology, Brno, Czech Republic pp 19–23

7. Krishans Z, Oleinikova I, Mutule A (2007) Planning for urban medium voltage network. In: 7th WSEAS/IASME International conference on electric power system, High Voltages, Electric machines (POWER'07), Venice, Italy (Conference Proceedings on CD 600–121)
8. Krishans Z, Neimane V, Anderson G (1999) Dynamic model for planning of reinforcement investments in distribution networks. In: 13th Power systems computation conference (Conference Proceedings), Trondheim, Norway,vol. 2. pp 863–869
9. Merkurjevs Y, Krishans Z, Oleinikova I, Mutule A (2007) Estimation methods of power system sufficient for 5-10 years horizon. Power Tech 2007 (Conference Proceedings on CD), Lausanne, Switzerland
10. Neimane V, Anderson G, Krishans Z (1998) Multi-Criteria Analysis Application in Distribution Network Planning. 33rd Universities Power Engineering Conference UPEC'98, (Conference Proceedings) Napier University, Edinburgh, UK, vol. 2 791–794

Chapter 9
The Database

Abstract Information technology database is related to system operating state database, which stores the latest information about a large technical system. Development management database data are used in a large technical system development process for technical analysis, economical analysis, optimization, and risk analysis. This chapter presents the structure of databases and operations with them. Development management technology database is divided into object (different large technical systems) databases. Each object database consists of system element parameters, transmission networks, electrical power plants, and research task. System elements are modelled according to the accuracy of output data and requirements of sustainable development management task. This chapter describes the basic principles of database structure and application Bakin et al. (Резервирование в энергосистемах и вопросы повышения надежности при глубоких ограничениях. 26–28, 1981); Dale et al. (Изв. АН СССР, 2:126–131, 1979); Krishans et al. (Вопросы надежности при эксплуатации и управления развитием энергосистем. Сб. Науч. Тр. НИИПТ. Энергоатомиздат, 21–24, 1986); Seiliger et al. (Вопросы построения автоматизированных информационных систем управления развитием электроэнергетических систем. Вып. 2. Структура и принципы построения 1. очереди АИСУ. 125–132, 1975); Seiliger et al. (Seminar on Comparison of Models of Planning and Operating Electric Power Systems (12) 1987)

9.1 Information Flow amongst Database and Software System for Management of Sustainable Development

Dynamic management of electric power system sustainable development can only be realized through special information technologies (IT). The structure of software system for management of sustainable development and informative links are illustrated in Fig. 9.1.

Fig. 9.1 Programs system
and informative links

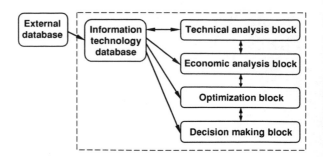

The IT under consideration, *LDM-TG'08*, makes it possible to review five levels of nominal voltage, e.g., 110, 220, 330, 400, and 750 kV. There is a database interconnected with an external database in the IT. The main IT blocks are as follows (see Fig. 9.1):

1. Technical analysis block, where major electrical parameters are calculated,
2. Economic analysis block which performs economic calculations,
3. Optimization block for optimal D-plan selection,
4. Decision-making block, where various forecasts related to load increase, interest rate, etc. are reviewed.

All blocks are database related. Economic analysis block is connected with technical analysis block; optimization block with economic analysis block and decision-making block.

The methods applied in block models depend on nominal voltage.

LDM-TG'08 is able to utilize its database and external database (see Fig. 9.1). If the external database is not available, the calculation can be performed using only information technology database. The external database provides possibility to enhance data accuracy along with saving time for data entry. In this chapter, database formation and structure are reviewed in detail, but in the forthcoming chapter, the structure of software system for management of sustainable development and its utilization schemes are provided.

9.2 Database Structure and Relevant Operations

External database structure. The external database is designed for data input of existing network from operational state calculation program. The external database

Fig. 9.2 External database
structure

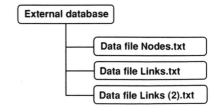

Fig. 9.3 Information technology structure

External database	**Data input**
	Nodes graphical input

utilization increases data accuracy and reduces the scope of work when creating new calculation objects of the network. The external database structure is depicted in Fig. 9.2.

Data on links in the external database differ from data which are utilized in the IT database, thus, special info model is required that prepares external database utilization in information technology.

Information technology structure. Module function is to adjust external database data on lines and loads to network object. Block structure is shown in Fig. 9.3.

Module *Data input* records external database information in database *External nodes* and *External links*.

Information on respective network object (portion) from the external database is sent to and stored in these three text files: *Nodes1.txt*, *Links.txt* (*VAR.txt*), and *Links(2).txt*.

Module *Nodes graphical profile input.* Nodes graphical profile input is done from the IT management window. After connecting to the module, a window appears where nodes dispositions are marked. From the list, user selects a node and places it in the scheme, indicating node location with cursor.

To draw external network scheme, all external node locations must be fixed. The external links appear automatically after initial and end nodes disposition. In order to perform calculations it is required, with the help of the IT database tools, to determine locations of feeding nodes and transformers. All calculations can be performed without load nodes disposition. In such case, the external network will

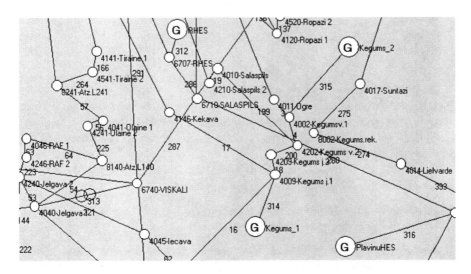

Fig. 9.4 Latvia's 110 kV network fragment after all external nodes disposition

Fig. 9.5 Object archive
structure

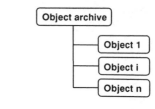

Fig. 9.6 Calculation object
structure

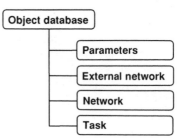

not be available, but still, the results of electrical calculations in the external network can be filed in the form of a table. Respectively, in order to chart external network calculation on the scheme it is required to locate load nodes (see Fig. 9.4).

Any external nodes can be interconnected with the IT links (perspective or existing).

Information technology database structure and operations with it. To facilitate database formation and calculations, the database is structured into calculation objects—various systems or one system parts. Calculation objects are stored in object archive. Object archive structure is shown in Fig. 9.5. Each object *i* has structure depicted in Fig. 9.6.

Operations. To work with object archive, the following operations are used:

1. Activate,
2. Attach,
3. Exclude,
4. Copy.

Let us review these operations in more detail.

Activate. This operation enables one to select previously created network object from network archive to form this object database (if it has not been created yet), as well as to adjust and supplement database to perform calculations (if the database is already formed).

Attach. This operation allows one to create data in the archive for new network object calculation. There are two possible sub-cases:

1. To create absolutely new network object or
2. To attach network object formed before.

Forming new network object, base parameters are formed automatically. Note that the network, D-actions, and conditions are formed for each object individually.

Fig. 9.7 Structure of block
Parameter catalogues

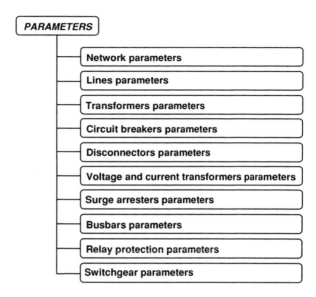

Exclude. This operation excludes unnecessary network calculation object from data archive, for example, operational object which has been already utilized for various D-actions analysis.

Copy. This operation enables copying the previously formed object. Creating a new operation object, the D-actions comprising one or more archive objects can be analysed.

9.3 Block *Parameter Catalogues*: The Significance and Structure

Parameter catalogues serve as manuals, as well as facilitate working with the data, reduce operations number, and allow one to standardize and relate to other database. The structure of parameter catalogue is shown in Fig. 9.7.

Parameter catalogues facilitate system development sustainability management process, setting parameters for links, nodes, and distribution units.

9.4 Block Network: The Structure and Operation with Network Scheme

The structure of information used in forming network model is depicted in Fig. 9.8.

Nodes are classified into feeding (B), loading (S), and generator (G) nodes. Nodes are characterized by information on rated voltage, node designation, and node type (B, S, and G).

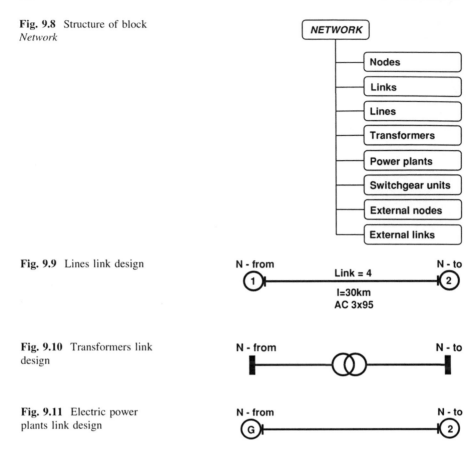

Fig. 9.8 Structure of block *Network*

Fig. 9.9 Lines link design

Fig. 9.10 Transformers link design

Fig. 9.11 Electric power plants link design

Links are classified into lines, transformers, and electric power plant links. Line links can be of only one type and one mark (see Fig. 9.9).

Information about link length, type, and respective wire mark is recorded in database *Lines*. Information about line links supplements database *Lines parameters*.

Transformers link connects two various rated voltage nodes (see Fig. 9.10).

Transformer initial node (*N-from*) must have higher rated voltage. Information about transformers is recorded in database *Transformers*.

Electric power plant link connects generator node to load node (see Fig. 9.11).

Electric power plants input window is shown in Fig. 9.14.

9.5 Object Input

Object input and database formation is performed concurrently. Illustration formation must be started with nodes disposition (drawing). The following icons shall be utilized (depending on nodes type):

- loads node,
- busbars, horizontal,
- cable network node,
- busbars, vertical,
- generator node.

With the cursor on the screen, node disposition is indicated. On the screen, window *Nodes* appear where information about nodes must be entered (see Fig. 9.12).

When input of at least two nodes is done, the links formation can be started.

Simple link is drawn with icon [/]. In links input window, information about link is entered.

Compound link is drawn with icon [⌐]. Compound link consists of several selected breaking points and line segments. First, initial links and end links are selected. Then links are drawn. Finally, link window information about link is entered.

If there are equal nominal voltages of initial node and end node of link, line links input window appears (see Fig. 9.13). If the initial node of link is a generator

Fig. 9.12 Nodes data input window type (*B/L/G* base/ loads/generator)

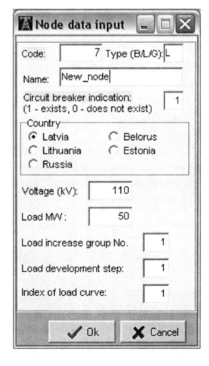

Fig. 9.13 Lines link input
window. *C* cable line,
G overhead line, *A* aerial self-
supporting cable

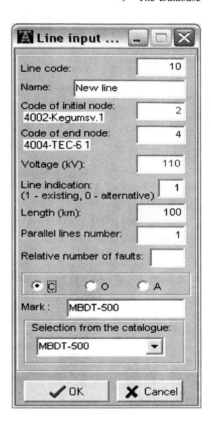

node then the link in the power plant and electric power plant link input window
appears (see Fig. 9.14).

To input information on switchgear, one must click on an icon ?⚡ and in the
scheme indicate with cursor a node where the switchgear is to be placed (see
Fig. 9.15).

Node code is formed automatically. Switching time per one disconnector must
be entered in units—hours. After that, switchgear must be selected from the catalogue (switchgear scheme appears automatically) as well as critical devices
according to input window.

9.6 Block Task: The Significance and Structure

The block of object database *Task* (see Fig. 9.6) contains information that is
required to form D-plans, estimate their technical, economic, and ecological criteria and select an optimal D-plan, observing perspective data ambiguous nature
and development of dynamic optimization area.

Fig. 9.14 Electric power
plant link input window

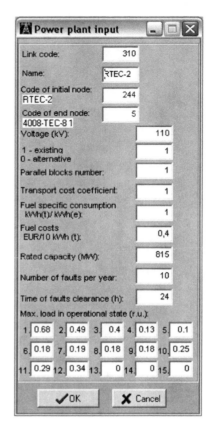

To perform the aforementioned functions, system development model is applied (see Sect. 1.3). The basic notions of the model are as follows: estimation period, decision-making period, D-step, and D-action.

The essential concept, applying network economic analysis complex is possible (alternative) D-actions—construction, reconstruction or dismantling of electrical power plants, lines or substations. D-actions are assigned with capital investments and network-related elements (elements, which as a result of D-action are included or excluded from calculation scheme). The user (users group) sets the actions according to the particular task, experience, etc. The IT actions, assigned by the user, are utilized as "bricks" for D-plan formation and optimal D-plan synthesis. The D-actions are electric systems development dynamic optimization task options (variables). All alternative D-actions set constitute this task optimization area. Examples of D-actions can be seen in Fig. 9.16.

The D-actions are classified as simple and compound. The compound D-action is formed from several simple D-actions utilizing logical conditions: "one from set" and "after".

With the help of generator element, energy saving technologies, as well as electric energy import from neighbouring power systems can be modelled. Solving the task concurrently on energy source and network development, annual load and

Fig. 9.15 Switchgear input
window

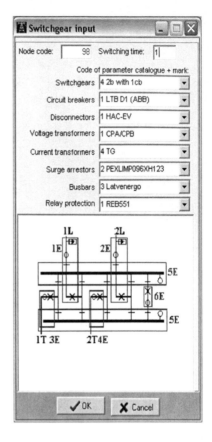

Fig. 9.16 Examples of
D-actions types. *1* Disconnection, *2* Switching on,
3 Line construction,
4 Substation construction,
5 Line and substation
construction

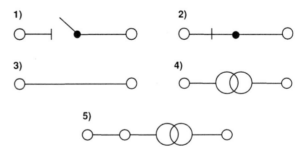

generation schedule are modelled with several operational states, for example,
winter maximum, winter night maximum, pre-peak operational state of power
system incorporating HPP, when HPP is tripped for operation and others.

The term *D-action realization* denotes completion of respective object construction, reconstruction, or dismantling (commissioning or decommissioning), but
not the time of construction, reconstruction, or liquidation. In our model it is

Fig. 9.17 Structure of block
Task

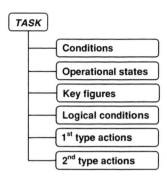

assumed that D-action is realized at the beginning of the first year of D-step. Time period of construction, reconstruction, or dismantling (liquidation) ends within this year. This time interval duration is observed when calculating respective D-actions costs (capital investments). Realizing D-action, calculation scheme is being changed.

Probable realization period. The beginning of this period is usually determined by time duration of object construction, reconstruction, or liquidation. In general, its end coincides with estimation period expiry, but in case of specific conditions the possible realization period may expire sooner—if, for example, on land site, where new substation construction is envisaged, other constructions can be initiated prior to calculation period expiry. The structure of block *Task* is depicted in Fig. 9.17.

As can be seen from Fig. 9.17, block *Task* is of complex structure. It comprises six parts. Let us review these part structures in more detail.

Conditions. Development conditions are information about D-steps (D-step period is one or more years, load increase, information on repair expenditures, etc.).

Year	D-step start calendar year;
Pi1	Load increase coefficient for Group 1 nodes (Group 1), r.u.;
Pi2	Load increase coefficient for Group 2 nodes (Group 2), r.u.;
Pi3	Load increase coefficient for Group 3 nodes (Group 3), r.u.;
Pi4	Load increase coefficient for Group 4 nodes (Group 4), r.u.;
Pi5	Load increase coefficient for Group 5 nodes (Group 5), r.u.;
Cut1	Code for the first line, which is cut in the last—emergency operational state (if zero is input, disconnection does not occur);
Cut2	Code for the second line, which is cut in the last emergency operational state (if zero is input, disconnection does not occur);
Cen	Energy price, EUR/kWh;
Cp	Capacity maximum price, EUR/kW/per year;
Cnep.	Electric energy non-delivery specific costs, EUR/kWh;
Crem.	Average costs of one fault clearance, r.u.;
Kmax	Admissible capital investments in D-step, TEUR;
Tav	Duration of emergency operational state, hours/year.

Operational states. Note that number of operational states must not exceed 15!

Reg.No.   Operational state index (ordinal number);
Name   Operational state name, 10 symbols;
Tre   Operational state duration, hours per year;
P1   1st type load with load schedule No. 1, r.u.;
P2   2nd type load with load schedule No. 2, r.u.;
P3   3rd type load with load schedule No. 3, r.u.;
P4   4th type load with load schedule No. 4, r.u.;
P5   5th type load with load schedule No. 5, r.u.

The maximal number of load schedules is five.

P1, P2, P3, P4, and P5 are utilized for nodes formation when load schedule number
   must be indicated, namely
P1 is the 1st load schedule number,
P2 is the 2nd load schedule number,
....
P5 is the 5th load schedule number.

If there are electric power plants in the object, the number of operational states
is 12 and duration is 730 hours. If emergency operational state is also reviewed,
the number of operational states is 13.

Key figures. The key figures relate to the entire calculation object (i.e., to all its
elements and D-steps) and may differ for various calculation objects. The key
figures are classified into two groups:

1. Economic indicators (interest rates, incomes, and decision-making period) and
2. Technical indicators (average cos φ, applied for calculation of apparent and
   reactive power).

Comparing various D-plans (searching for optimal D-plan), the key figures
for all D-plans must be equal. Key figures are changed performing sensitivity
analysis.

The input of key figures only for the first year is utilized in the calculation.
Thus, the input of key figures for the rest of the years is not required.

int %   Interest (credit) rate, %;
T decision   Decision-making period in years (six years by default);
Disk. %   Discount %;
I Cm. %   Network maintenance costs inflation %;
I fuel. %   Fuel inflation %;
I invest. %   Capital investments inflation %;
Incom.   Income from electric energy unit sales, EUR/kWh;
cos (*f*)   Consumers average cos φ.

Table 9.1 Key figures for the 1st type D-actions

Key figures	Symbols in IT
Index of D-action	A. code
Name of D-action, 14 symbols	Name
Capital investments (to relevant link), TEUR	K
Materials and equipment costs, TEUR	Km
Construction and assembling costs, TEUR	Kc
Annual allocations for depreciation in % of capital investments	p
Annual allocations for maintenance in % of capital investments	pu
Maintenance costs—operational and annual costs, in disproportion to capital investments and irrespective of load (land rent, taxes, incentive levies on environment protection as well as for repairs in case of emergencies, etc. depending on the task), TEUR/year	C
Code of links relevant to the D-action	L.code
Indication: if construction link is envisaged, the mark "+" is placed; if liquidation link is envisaged, the mark "−"is placed	±
Initial year of D-action probable realization period; if in the stated task D-action realization is not possible, then it must be fixed as Tinit = 0	Tinit.
End year of D-action possible realization period; if D-action realization end year coincides with estimation period, then it can be recorded as Tend = 0	Tend
D-action number, prior to which the stated D-action must not be realized (if there is no such limitation, then it is fixed as after = 0)	after

Logical conditions. Logical conditions are characterized by D-action set. It determines that none of D-plans and none of D-steps may contain two or more elements of the set. For example, if several conductors' trade marks are considered for the stated line, then only one mark of conductor (wire) may be utilized concurrently in the line.

The set is attributed as D-action code, *A. code*, list. Each set is recorded in one line with no spaces. The maximal number of D-actions is 10 and the maximal number of sets is 10.

The 1st type D-actions. The 1st type D-actions are modelled for removing subsistent links and constructing new ones. Removable links may be external links or information technology links, but new constructed links may only be information technology links. Note that for the 1st and 2nd D-actions there is co-numeration; it must not coincide (Table 9.1).

If the number of links in an action is higher than one, then there will be recorded as many lines, as links in the D-action; and the second and further lines are filled up as follows: *A.code* - repeat, *Name* - remain blank or write in the comments on the stated line, further on write information on the next link.

The 2nd type D-actions. The 2nd type D-actions are modelling such complex D-actions as replacement of wires in the subsistent lines, transformers replacement, and transfer from overhead lines to cables. The 2nd type D-actions may also be applied if the existing link is an external link (Table 9.2).

Table 9.2 Key figures of the 2nd type D-actions

Key figures	Symbols in IT
Index of D-action	A. code
Name of D-action, 14 symbols	Name
Capital investments (for relevant link), TEUR	K
Materials and equipment costs, TEUR	ME
Construction and assembling costs, TEUR	CA
Annual allocation for depreciation in % of capital investments	p
Annual allocations for maintenance in % of capital investments	pu
Maintenance costs—operational and annual costs, in disproportion to capital investments and irrespective of load (land rent, taxes, incentive levies on environment protection, as well as for repair in case of emergencies, etc. in relation to the task), TEUR/year	C
Code of links relevant to the D-action	L.Code
A new mark of relevant links (lines or transformer) from the respective catalogue	Mark
Initial year of D-action probable realization period; if in the stated task D-action realization is not possible, then it must be fixed as Tinit = 0	Tinit.
End year of D-action possible realization period; if D-action realization end year coincides with estimation period, then it can be recorded as Tend = 0	Tend
D-action number, prior to which the stated D-action must not be realized (if there is no such limitation, then it is fixed as after = 0)	after

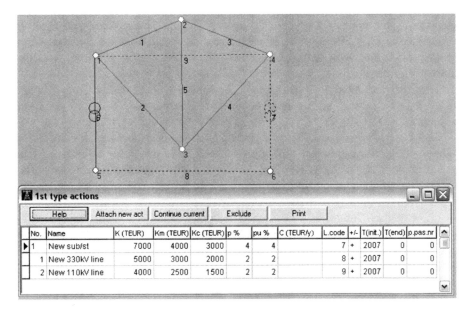

Fig. 9.18 Example of the 1st type D-actions formation (for symbols, see Table 9.1) *No.* index of D-action

Let us review 1st type D-actions formation example. In Fig. 9.18 calculation example is given on 330 and 110 kV network scheme. In the example, two D-actions are formed.

External nodes are: 4001, 4002, 4003 and 4004

External links are: LN-101, LN-102, LN-103, LN-104 and LN-105

Information technology nodes are: 5 (Ap/st1) and 6 (Ap/st2) (see Fig. 9.18). Node sub/st1 is feeding node (B type node).

D-action 1. Information technology links include transformer link 6, transformer link 7, and 330 kV lines 8 and 9 (see Fig. 9.18). Here, substation 1 is existing, but substation 2 and 330 kV line 3 are alternatives, the construction of which would improve power supply reliability and reduce energy losses in 110 kV network. Figure 9.18 also presents action formation window, *1st type D-actions.* D-actions are recorded in two rows: in the first row new substations are recorded, but in the second one new 330 kV lines.

D-action 2. D-action 2 is an alternative to D-action 1. In node 6 (sub/st2) substation is not built, instead, a new 110 kV line 9 between nodes 1 and 4 will be built. Link 9 is information technology link, connecting two external nodes. The 2nd action is recorded in D-action formation window: *1st type D-actions* in the third row.

References

1. Babkin VF, Itkin JA, Krishans ZP, Dale VA (1981) Optimization model structure of large power system transmission networks development taking into account reliability. In: Резервирование в энергосистемах и вопросы повышения надежности при глубоких ограничениях. Frunze (Bishkek), Kyrgyzstan, 26–28 (in Russian)
2. Dale VA, Krishans ZP, Paegle OG, Greivule IJ (1979) Software complex for analysis of power system transmission networks development. Изв. АН СССР, 2:126–131 (in Russian)
3. Krishans ZP, Abramova HJ, Zaslavskaja TB, Abramenkova NA, Gelfand VI (1986) Power system transmissionnetworks optimization taking into account static stability. In: Вопросы надежности при эксплуатации и управления развитием энергосистем. Сб. Науч. Тр. НИИПТ. Энергоатомиздат, Leningrad, Russia, 21–24 (in Russian)
4. Seiliger AN, Kusnetsova ON, Malkin PA, Valevicene JTP, Dale VA, Krishans ZP, Paegle OG (1975) Information technology for technical economic calculations of power system transmission networks. In: Вопросы построения автоматизированных информационных систем управления развитием электроэнергетических систем. Вып. 2. Структура и принципы построения 1. очереди АИСУ. Irkutsk, Russia, 125–132 (in Russian)
5. Seiliger AN, Krishans ZP, Kusnetsova ON, Lasebnik AN, Paegle OG, Tsallagova ON, Erhardt EI (1987) Use of optimizing and assessing models to analyze and determine the development of electricity networks in power systems. United Nations, Economic Commission for Europe. Seminar on Comparison of Models of Planning and Operating Electric Power Systems (12). Moscow, Russia

Chapter 10
Software System for Sustainable Development Management

Abstract LTS sustainable development management IT software system can be divided into three software complexes: (1) technical analysis complex, (2) economical analysis complex, and (3) development dynamic optimization complex. According to these complexes, Chap. 10 is subdivided into tree sections. Section 10.1 describes (1) formation technique of development state and operating state, (2) calculation of power flow and voltage, (3) calculation of switchgear reliability, (4) calculation of transmission network reliability, and (5) calculation of electric energy annual balance. Section 10.2 describes (1) formation of development plan, (2) development plan economic comparison, (3) economic criteria and their calculation algorithms, (4) sensitivity analysis of development plans. Section 10.3 describes *development dynamic optimization* complex's structure and functions: used optimization criteria, limitations, methods and algorithms. In the process of sustainable development management, optimization is performed in uncertainty condition. Therefore, in optimization complex, optimization result risk analysis is of great importance.

10.1 Software Complex for Technical Aspects Analysis

10.1.1 Significance of Block Technical Analysis in Sustainable Development Management Software System

The structure of software system for management of electric power system sustainable development is represented in Fig. 10.1.

Parameter analysis of the D-plan assigned by user is performed by assessment software. Optimization software determines an optimal D-plan automatically. Both assessment and optimization software are widely applied in practice.

Z. Krishans et al., *Dynamic Management of Sustainable Development*,   131
DOI: 10.1007/978-0-85729-062-5_10, © Springer-Verlag London Limited 2011

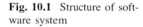

Fig. 10.1 Structure of software system

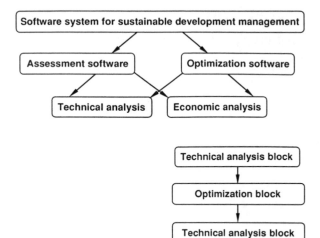

Fig. 10.2 Arrangement of software complex for technical aspects analysis

The IT comprises assessment and optimization software.

Technical aspects analysis software complex arrangement is illustrated in Fig. 10.2.

As shown in Fig. 10.2, technical analysis block is used in two locations, before and after optimization block. Besides, technical analysis block is also included in optimization block.

Technical analysis before optimization block is done to determine bottlenecks in electric network and form alternative D-actions focused on further development.

Technical analysis after optimization block is done to verify technical parameters of the determined D-plan. If technical parameters do not meet the requirements, the user forms supplementary D-actions and repeats optimization. Technical analysis is performed for the D-plan under consideration, D-step t and operational state re. To facilitate operation for user, the program contains D-plan formation block.

Menu *Development plans* is utilized if user is willing to assign comparable D-plans on his own. If optimization is performed, then in Windows, available in this selection, up to ten best possible D-plans will be formed. Since admissible number of D-actions is 30, D-plans formation (or consideration) is classified into two parts, including 15 D-actions in each (see Fig. 10.3).

In menu window on the left, D-action names from database *1st type D-actions* and *2nd type D-actions* are located. Its enumeration may not correspond to respective enumeration of the database. The main point is its name.

If D-action in the respective D-plan (1, 2,..., 10) is not realized, then it corresponds to - - - -, otherwise year of realization is indicated. If indicated year is not entered into database *Conditions*, then the nearest least calendar year, which is set in *Conditions*, is taken automatically. D-plans themselves differ from realized D-actions and years of realization. The first and the second D-plans realize one and the same D-action that means substation A construction, but in various years: the 1st D-plan in 2010 and the 2nd D-plan in 2012.

D-actions:	D-plans 1	2	3	4	5	6	7	8	9	10
1 Sub./st. A	2010	----	----	----	----	----	----	----	----	----
2 Line AB	2013	----	----	----	----	----	----	----	----	----
3 Line AC	2020	----	----	----	----	----	----	----	----	----
4 Sub./st. C	----	2012	----	----	----	----	----	----	----	----
5 Line CD	----	2015	----	----	----	----	----	----	----	----
6 Line CE	----	2020	----	----	----	----	----	----	----	----
7 Power station	----	----	2015	----	----	----	----	----	----	----
8		----	----	----	----	----	----	----	----	----
9		----	----	----	----	----	----	----	----	----
10		----	----	----	----	----	----	----	----	----

Fig. 10.3 Example of D-plans consideration and formation

Fig. 10.4 Structure of technical analysis block

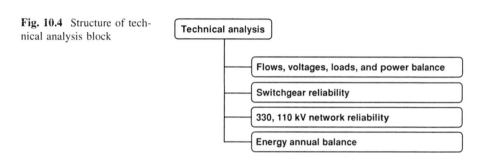

If for a certain D-plan, all D-actions correspond to - - - -, this D-plan conforms to existing network scheme. Selecting the existing D-plan, calculation on subsistent network scheme, which is considered as the last from the D-plans is performed automatically. The maximal number of formatting D-plans is 10.

The structure of technical analysis block is shown in Fig. 10.4.

10.1.2 Flows, Voltages, Loads and Operational State Summary Indicators (Power Balance, Number of Overloaded Lines)

Entering this sub-menu window, menu of electrical calculation conditions appears (see Fig. 10.5).

There is a possibility for user of setting his preferred calculation conditions:

1. D-plan,
2. D-step,
3. Operational state,
4. Load coefficient,
5. As well as calculation type:

Fig. 10.5 Formation of
electrical calculation
conditions

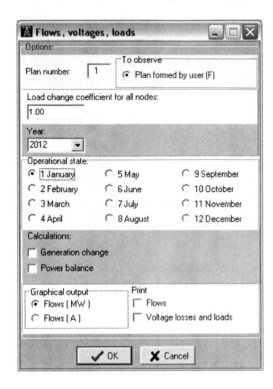

– Only flow calculation,
– Flow calculation for changed generator system operational state if generator
  parameters are changed Pnom, cdg,
– Load balance in operational state and overloaded lines number in power
  systems.

In the menu, the following is formed by default:

1. First D-plan,
2. First D-step,
3. Annual maximal operational state that, respectively, corresponds to state No. 1,
4. Load changing coefficient equal to 1, if calculation is performed for average
   monthly load and is equal to 1.3, if calculation is performed for the maximal load.

The results of flows, voltage losses and load calculations are forwarded to the
scheme or printed in paper format. The results can be obtained for any D-plan in
any D-step in a certain operational state.

The available scheme information contains:

• Name of nodes,
• Voltage losses in nodes (%),
• Loads in nodes (MW),
• Power flows in links (MW),
• Load in links (in related units).

Graphically highlighted are those voltage losses in a node that exceed admissible values as well as links whose load is higher than 1.0 and 1.2 (see Fig. 10.7).

10.1.3 Calculation of State with Changed Power Balance Generation System

If you are willing to perform calculation of a state with changed power balance generation system, then from information technology management window you must select *Electrical calculations → Flows, voltages, loads, balance*. Entering this menu, the window of electrical calculation conditions appears (see Fig. 10.6).

In the window, there is automatically formed the following: (1) the first D-plan, (2) load changes coefficient value for all nodes, equals 1, (3) the first D-step and (4) the first operational state. Now user can input the desired information such as (1) D-plan, (2) load changes coefficient value, (3) D-step and (4) operational state. Additionally, user must mark selection *Generation replacement*. Approving selection with OK button, the window of generation parameters appears on the screen (see Fig. 10.7), where user can change generation parameters for one or more electric power plants. Approving the selection with OK button, on the screen appears the window of the results of electrical calculations for the changed generation system.

Fig. 10.6 Generation parameter window

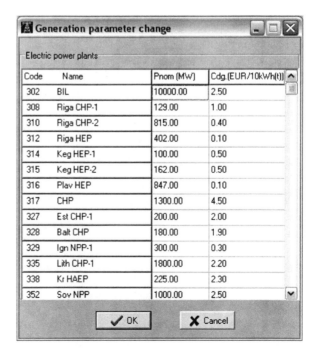

Code	Name	Pnom (MW)	Cdg.(EUR/10kWh(t))
302	BIL	10000.00	2.50
308	Riga CHP-1	129.00	1.00
310	Riga CHP-2	815.00	0.40
312	Riga HEP	402.00	0.10
314	Keg HEP-1	100.00	0.50
315	Keg HEP-2	162.00	0.50
316	Plav HEP	847.00	0.10
317	CHP	1300.00	4.50
327	Est CHP-1	200.00	2.00
328	Balt CHP	180.00	1.90
329	Ign NPP-1	300.00	0.30
335	Lith CHP-1	1800.00	2.20
338	Kr HAEP	225.00	2.30
352	Sov NPP	1000.00	2.50

Generation parameter change

Electric power plants

✓ OK   ✗ Cancel

Fig. 10.7 Structure of operational state summary indicators

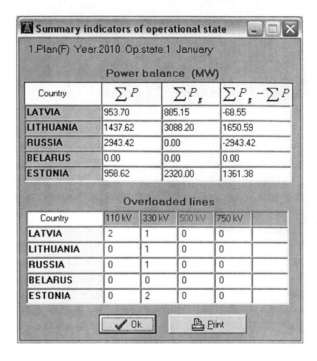

10.1.4 Calculation of Operational State Active Power Balance

To perform operational state active power balance calculation, one must select the following menu: *Electrical calculations → Flows, voltages, loads, power balance*. Entering this menu, the window of electrical calculation conditions appears (see Fig. 10.5).

In the window, the following are automatically formed: (1) the first D-plan, (2) load changes coefficient value for all nodes, equals 1, (3) the first D-step and (4) the first operational state. Now user can input the desired information such as (1) D-plan, (2) load changes coefficient value, (3) D-step and (4) operational state. Additionally, user must mark selection *Power balance calculation*. Approving the selection with OK button, on the screen appears the window of operational state summary indicators (see Fig. 10.7), with calculation results of power balance and number of overloaded lines.

10.1.5 Analysis of Operational State with Outages

In the process of transmission electric network and generation sustainable development management, the essential role is drawn to analyse operational states with outages.

Operational states with outages can be modelled in two ways:

1. At annual criteria level,
2. At electrical calculations level.

10.1.5.1 Analysis at Annual Criteria Level

For the analysis, menu *Operational states* and *Conditions* must be used. As the last, we add operational state with disconnected links, usually with maximal load and not long duration in hours.

In *Conditions* columns *Atsl.1* and/or *Atsl.2*, the recording of the link number which is disconnected due to fault or scheduled repair must be done. The maximal number of coincidentally disconnected lines is 2.

10.1.5.2 Analysis at Electrical Calculations Level

Lines disconnection with icon ✂ can be started when the calculation is performed for required (normal) operational state. After that, with cursor user selects the lines, which are drafted for disconnection (see Fig. 10.8). The maximal number of disconnected lines is 10.

Link for disconnection 4 is marked with sign "✗". For review of results in the window of electrical calculation conditions, the D-step for consideration must be indicated; besides, any operational state and D-plan can be selected. The results are illustrated in Fig. 10.9.

10.1.6 Switchgears Reliability

User can calculate switchgears reliability criteria for any D-plan, D-step, switch-gear and type of device. In order to change switchgear type and/or device type,

Fig. 10.8 Network scheme

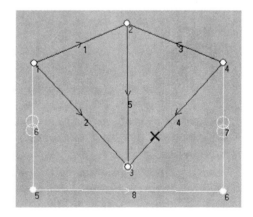

Fig. 10.9 Calculation results
of network with disconnected
link 4

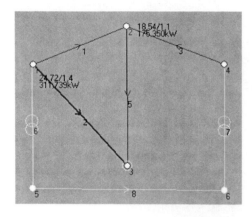

Fig. 10.10 Conditions of
switchgear reliability criteria
calculation

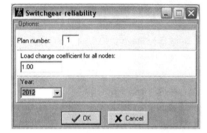

user must press icon and in the scheme with cursor indicate a node where
switchgear is located. The window of switchgear data input then appears on the
screen (see Fig. 9.15). Calculations are performed using menu *Electrical calculations → Switchgear reliability* or icon ▢.

Selecting this menu, the window of switchgear reliability criteria calculation
conditions appears (see Fig. 10.10). Approving selection with OK button, on the
screen appears the calculation results—switchgear reliability criteria (see
Fig. 10.11).

10.1.7 330, 110 kV Network Reliability

330, 110 kV network reliability calculations can be performed by user for any
D-plan in any D-step. For calculation of network reliability, all existing and
anticipated switchgears (may be simplified, without transformers) must be entered.
Calculations are performed using menu *Electrical calculations → 330, 110 kV
network reliability* or icon ⊞.

Entering this menu, the window of condition selection for 330, 110 kV network
reliability criteria calculation opens (see Fig. 10.12). In this window, these conditions are formed by default: (1) first D-plan, (2) first D-step and (3) load changes

Fig. 10.11 Switchgear
reliability criteria

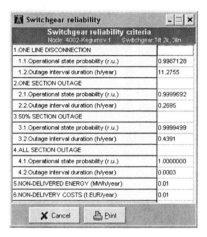

Fig. 10.12 Conditions for
330, 110 kV network criteria
calculation

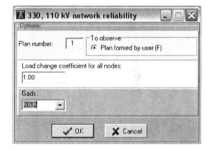

coefficient value for all nodes, equal to 1. User may also enter input information of
this choice like (1) D-plan, (2) D-step and (3) load changes coefficient. Approving
selection with OK button, on the screen one can see the results—criteria of 330,
110 kV network reliability (see Fig. 10.13).

10.2 Software Complex for Economic Aspects Analysis

10.2.1 Structure of Complex, Components Function

The structure of economic aspects is shown in Fig. 10.14.

The complex of economic analysis consists of three blocks.

Block *Development plans comparison analysis* calculates technical economic
indicators for all formed D-plans, observing conditions specified in the database.

Blocks *Development plans sensitivity analysis* and *Risk analysis* calculate
technical economic indicators for each D-plan, observing various condition forecasts, allowing making decisions under information uncertainty conditions.

Fig. 10.13 330, 110 kV
network reliability criteria

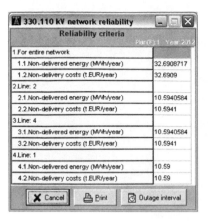

Fig. 10.14 Structure of
economic analysis complex

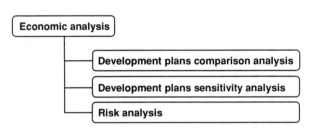

The functions of the complex are as follows:

1. D-plans formation;
2. D-plans technical analysis;
3. D-plans economic comparison;
4. Decision making under information uncertainty conditions.

The sequence of operation is the following:

1. Formation of D-plans for comparison.
   Utilizing development actions, user himself is forming D-plans for comparison.
2. D-plans technical analysis that ensures technical validity of all D-plans within
   the entire estimation period.
   If required, user creates new D-actions and forms new D-plans to provide
   technical validity of all D-plans;
3. Economic criteria calculation for all technically valid D-plans.
   The main criterion is total discounted incomes within estimation period:

$$C_s = \sum_{t=1}^{t=tm} C_t \cdot \text{Inf}_t \cdot d_t, \qquad (10.1)$$

where
C_s—total discounted income/expenses
t—D-step

Fig. 10.15 Utilization sequence of D-plans analysis and decision making complex

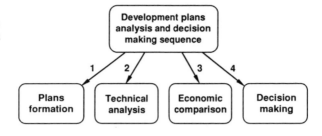

Fig. 10.16 Structure of block *Economic comparison*; *asterisk* for D-plans formed by user; *double asterisks* optimization software for formed D-plans

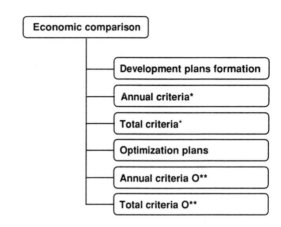

tm–estimation period

C_t—income/expenses in D-step t

Inf_t—inflation rate in step t

D—discounting coefficient

Income part of C_t considers income from sold energy, but expenses part—service and maintenance costs, capital investments, losses cost, taxes and penalty function for non-compliance with technical limitations. In case of unambiguous information, user selects the best D-plan according to this criterion.

4. Special analysis in case of information uncertainty.

For technical analysis, direct and iteration flow calculations are applied, as well as division point optimization methods and electric power plant optimization methods.

For decision making under information uncertainty, risk analysis and sensitivity analysis are employed.

Utilization scheme of D-plans analysis and decision making complex In order to realize the mentioned functions (operations), various blocks are applied. Utilization sequence of complex utilization is indicated in Fig. 10.15.

Operations with these blocks are performed from information technology main menu window. The structure of block *Technical analysis* is represented in Fig. 10.4, and the structure of block *Economic comparison* in Fig. 10.16.

Formation function of block *Economic comparison* is to enable the user to input D-plans which he intends to compare using D-actions. The data are entered in D-plans matrix (see Fig. 10.3).

10.2.2 Block Technical Economic Analysis of Plans Assigned by User: Calculation Results and Its Output File

Technical economic analysis of D-plans assigned by user is performed on consideration of all assigned D-plans in consecutive order (option *Development plans*).

In Fig. 10.17, D-plans formation occurs in block *Plans model*. All D-steps of each D-plan are considered starting with the first block *Development process model*. At the beginning of each D-step, network scheme is formed, utilizing D-actions. The core of technical economic analysis block of D-plans assigned by user is *Development step model*. In *Development step model*, power balance

Fig. 10.17 Flowchart of D-plans technical economic analysis. *i* index (ordinal number) of D-plan, *t* current D-step, *TM* number of D-steps, *m* number of D-plans

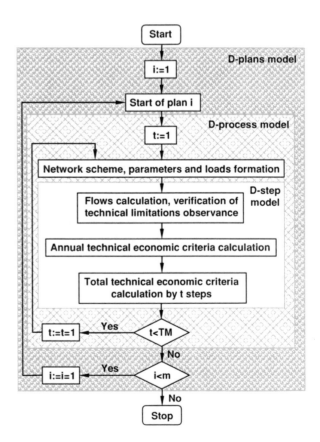

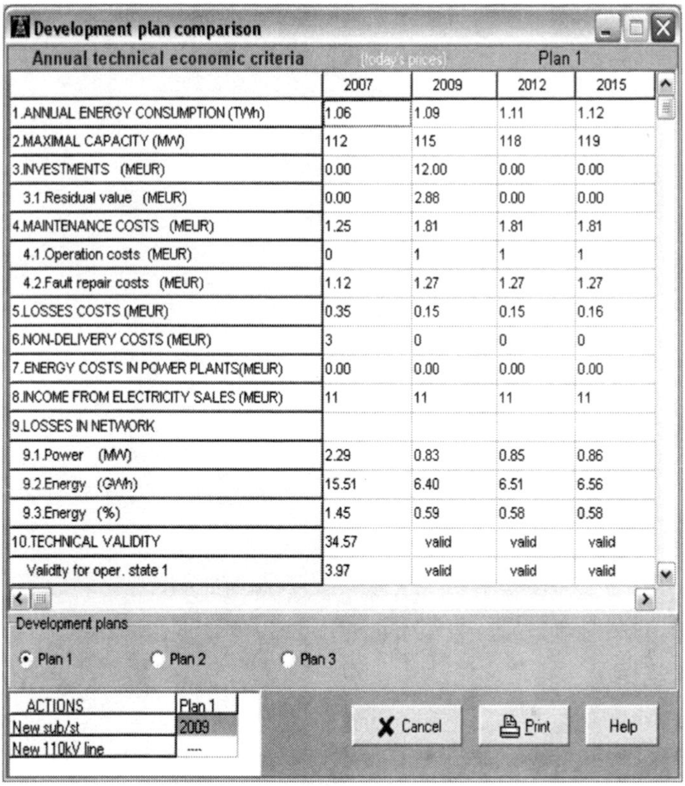

Development plan comparison

Annual technical economic criteria (today's prices) Plan 1

	2007	2009	2012	2015
1.ANNUAL ENERGY CONSUMPTION (TWh)	1.06	1.09	1.11	1.12
2.MAXIMAL CAPACITY (MW)	112	115	118	119
3.INVESTMENTS (MEUR)	0.00	12.00	0.00	0.00
3.1.Residual value (MEUR)	0.00	2.88	0.00	0.00
4.MAINTENANCE COSTS (MEUR)	1.25	1.81	1.81	1.81
4.1.Operation costs (MEUR)	0	1	1	1
4.2.Fault repair costs (MEUR)	1.12	1.27	1.27	1.27
5.LOSSES COSTS (MEUR)	0.35	0.15	0.15	0.16
6.NON-DELIVERY COSTS (MEUR)	3	0	0	0
7.ENERGY COSTS IN POWER PLANTS(MEUR)	0.00	0.00	0.00	0.00
8.INCOME FROM ELECTRICITY SALES (MEUR)	11	11	11	11
9.LOSSES IN NETWORK				
9.1.Power (MW)	2.29	0.83	0.85	0.86
9.2.Energy (GWh)	15.51	6.40	6.51	6.56
9.3.Energy (%)	1.45	0.59	0.58	0.58
10.TECHNICAL VALIDITY	34.57	valid	valid	valid
Validity for oper. state 1	3.97	valid	valid	valid

Development plans

Plan 1 Plan 2 Plan 3

ACTIONS	Plan 1
New sub/st	2009
New 110kV line	----

Cancel Print Help

Fig. 10.18 Annual technical economic criteria

calculation and verification of technical limitations observance occurs. If technical limitations are not observed, then fuzzy constraint method (see Chap. 5) is applied to calculate penalty functions. Further on, in *Development step model* calculation of *Annual technical economic criteria* and *Total technical economic criteria* is performed.

As a result of analysis, annual technical economic criteria are calculated for each D-plan and each D-step. An example of *Annual technical economic criteria* is shown in Fig. 10.18.

D-Plans economic analysis is performed with the menus *Annual criteria* and *Total criteria*.

Annual expenses are shown for the first year of each D-step of Fig. 10.18 which depicts economic and technical criteria in table form. One table contains annual technical economic indicators for only one of the D-plans; to review other D-plans, options given under the table must be selected.

In the table of total technical economic criteria (see Fig. 10.19), incomes are marked with "+" symbol, but expenses with "−" symbol. In the second part of the table, D-plans comparison with optimal D-plan is provided. Technical validity is

Fig. 10.19 Total technical economic criteria

also indicated in the table. "Valid" means that within the entire estimation period all technical limitations are observed; "non-valid" technical limitations are not observed. If there are non-valid D-plans, it is required to make technical analysis and add D-actions, as well as to perform re-optimization.

Note that investment repayment schedule may be reviewed for any D-plan.

10.2.3 Block Plans Sensitivity Analysis

D-Plan sensitivity analysis block enables one to analyze the influence of one specific factor on D-plans correlative effectiveness under information uncertainty conditions. Sensitivity analysis block helps to determine the following:

1. Essential and non-essential factors in economic estimation of D-plans,
2. Essential factors boundary values when relative estimation is changed (new optimal D-plan appears).

D-Plans sensitivity analysis flowchart is shown in Fig. 10.20. D-Plans sensitivity can be analyzed depending on various researchable factors. Researchable factors are depicted in Fig. 10.21. Let us review in detail, sensitivity analysis with regard to capital investments. First, let us select the appropriate factor (see Fig. 10.21).

Fig. 10.20 D-Plans sensitivity analysis flowchart; *j* step of researchable factor datiq change

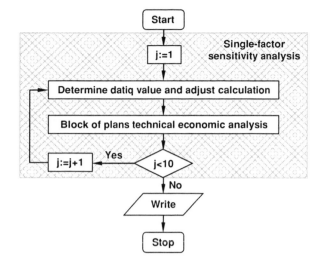

Fig. 10.21 Menu for selection of sensitivity analysis factors

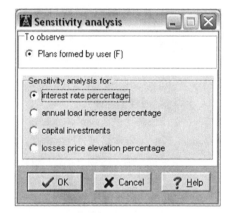

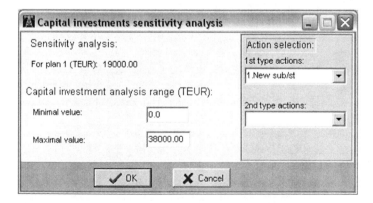

Fig. 10.22 Menu for selection of capital investments analysis range

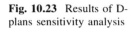

Fig. 10.23 Results of D-plans sensitivity analysis

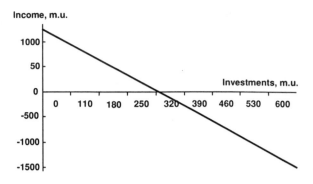

Further on, the selected range of factor changes is entered (see Fig. 10.22).

The range is divided into eight equal intervals. In each interval, all criteria calculation is repeated for each D-plan. The corresponding parameters in the database are not changed. The result is represented by graphical chart (see Fig. 10.23), which makes it possible to rapidly estimate specific parameters uncertainty influence on D-plans selection. It is worth noting that sensitivity analysis is only relevant for technically valid D-plans.

10.3 Complex of Software for Dynamic Development Optimization

10.3.1 Functions of Optimization Complex, Optimization Methods and External Links

The major function of *Optimization complex* is optimal D-plan selection. Based on D-actions formed by the user and database information about network object and its possible development conditions, *Optimization complex* determines when and which actions must be realized. As a result of optimization, not only optimal D-plan is determined, but also competitive D-plans (up to ten D-plans), if there are such.

In *Optimization complex*, the OIS method (see Chap. 5) and the maximal effect function OIS searching method are used (see Chap. 6), elaborated by the Laboratory for Power Systems Mathematical Modelling (PSMM) of Institute of Physical Energetics.

Optimization criterion is total discounted income within the estimation period (see formula 10.1). Income part of C_t considers income from sales of energy, but expenses part includes service and maintenance costs, capital investments, losses cost, taxes and penalty function. Optimization occurs by observing admissible transformer and lines loads and admissible voltage losses. For this purpose, fuzzy constraint method is utilized (penalty functions).

Fig. 10.24 Information links of optimization complex

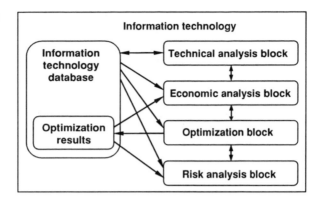

Fig. 10.25 Structure of *Optimization block*

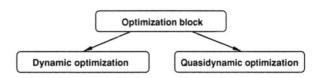

Optimization complex information links are illustrated in Fig. 10.24.

Optimization results (ten best D-plans) are sent to the database section *Optimization results*. Further on this can be processed in *Economic analysis block* (see Sect. 10.2), and in case of information uncertainty, also in *Risk analysis block* (see Sect. 10.3.3).

10.3.2 Structure of Optimization Block

The structure of *Optimization block* is represented in Fig. 10.25.

As shown in Fig. 10.25, *Optimization block* consists of two sub-blocks. Selecting option *Optimization* in the menu, sub-menu appears for selection, *Dynamic optimization* and *Quasidynamic optimization*. The functions of sub-selections are considered below.

10.3.3 Sub-block Dynamic Optimization

Dynamic optimization is performed for a certain decision making period. In each dynamic optimization step (up to 25 steps), all competitive D-plans are remained (up to 200). As a result of selection, best D-plans are automatically formed, which are obtained through optimization (10 D-plans), so both technical and economic analysis may be performed for these D-plans.

Dynamic optimization task can be formulated as follows: to determine optimal D-plan opt G, that has the maximal value for criterion

$$F(\text{opt } G) = \max_{G \in \{G\}} \sum_{t=1}^{TM} F(t, e(t)), \tag{10.2}$$

where

    TM—number of D-steps;

    G—D-plan;

    {G}—set of all possible D-plans;

    $e(t)$—D-state in step t;

    $g(t,e(t))$—objective function part in D-step t

    D-plan is described as states sequence like this:

$$G = e(1) \subseteq e(2)\ldots e(t-1) \subseteq e(t)\ldots e(T-1) \subseteq e(T). \tag{10.3}$$

D-states in step t can be expressed by a set of realized D-actions (see Fig. 10.26). In Fig. 10.26, the task is reviewed with two D-actions; possible D-states are:

    {0}—none of D-actions is realized;

    {1}—1st D-action is realized;

    {2}—2nd D-action is realized;

    {1,2}—1st and 2nd D-actions are realized

Thus, it must be clarified which D-actions and in which D-step they must be realized.

Symbol $e(t-1) \subseteq e(t)$ expresses the meaning that set $e(t)$ contains all set $e(t-1)$ D-actions. Possible D-plans set can be illustrated with development graphic.

Dynamic development optimization software complex can be illustrated with a flowchart (see Fig. 10.27).

Fig. 10.26 Development graph

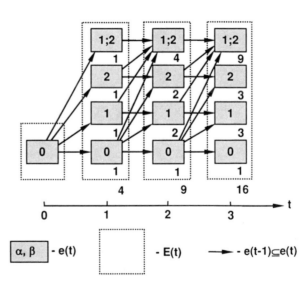

α, β - e(t)   - E(t)   - e(t-1)⊆e(t)

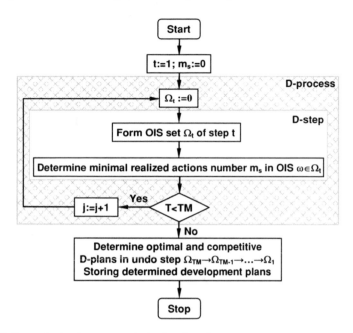

Fig. 10.27 Dynamic development optimization flowchart

Optimization with the help of dynamic programming is done in the following way. Optimization is a TM-step process, in which all steps are reviewed in consecutive order $t = 1, 2,...,$ TM and in each step, solutions are determined by t steps (made decisions). The process is not dependent on steps number. Decisions in step t are obtained from step $t - 1$ made decisions by applying recursive equation:

$$f(t, e) = F(t, e) + \max_{\{e(t-1) \subseteq e\}} f(t - 1, e(t - 1)). \tag{10.4}$$

Functional $f(t,e)$ expresses the minimal objective function by t steps to get from the initial state $e = 0$ in step $t = 0$ to state e in step t.

Optimization is performed by OIS method (see Chap. 5) which, as compared to dynamic programming, allows one to considerably enlarge the number of alternative D-actions. According to OIS method, in step t only OIS for step $t + 1$ are searched and stored. OIS set Ω searching is a multi-step process, the flowchart of which is shown in Fig. 10.28.

The core of OIS formation block is block *State analysis*. In this block, D-state $e(t)$, which is obtained in previous block by merging state $e(m - 1)$ with D-action j, is analyzed.

Let us consider an example.

Assume that D-actions number is 20, but realized D-actions number is 4; $e(m - 1) = \{1,5,8\}$;

if $j = 10$, then $e(t) = \{1,5,8,10\}$ with four realized D-actions $m(e + (t))$;

if $j = 5$, then $e(t) = \{1,5,8\}$ with three realized D-actions $m(e + (t)) \neq m$ D-state is non-valid and is not analyzed.

Fig. 10.28 Flowchart of OIS formation block. α-mark: $\alpha = 0$, in step m not any of OIS is found; $\alpha = 1$, in step m at least one optimal initial state is found; i D-state $e(m - 1)_i$ index (ordinal number); j D-action index (ordinal number); $m(e(t))$ number of realized D-actions in D-state $e(t)$

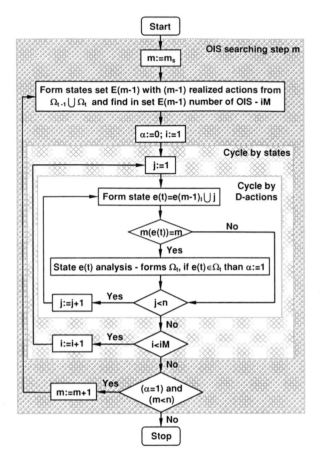

D-state analysis block performs the following operations:

1. Verification of logical limitations;
2. Power flow, voltage losses and reliability criterion calculation;
3. Objective function part in D-step t calculation;
4. Functional $f(t,e)$ calculation;
5. Verification whether $e(t)$ belongs to OIS.

Let us review the above operations in more detail.

1st operation Major logical limitations are: (a) only one from set "or" (b) "after". Let us review some examples.

Limitation "or" For example, wire cross-section selection for line 3xA50 or 3xA70. Let us assume that 3xA50 corresponds to D-action 8, but 3xA70 to D-action 10, then D-state $e(t) = \{1,5,8,10\}$ does not satisfy this limitation as in this state D-actions 8 and 10 are concurrent, and hence $e(t)$ is not analyzed further.

Limitation "after" For example, D-action 3 corresponds to construction of substation A, but D-action 10 to transformer replacement in substation A. Then

D-state $e(t) = \{1,3,8,11\}$ is valid, but D-state $e(t) = \{1,5,8,11\}$ is non-valid, as there is no D-action 3.

2nd operation is the same as technical analysis block (see Sect. 10.1), only in optimization process; the results for separate D-states are with no output but only utilized in the 3rd operation for objective function calculation.

3rd operation Objective function component in D-step t for D-state $e(t)$ is calculated as follows:

$$g(t,e(t)) = (C_{\text{Inc}}(t,e(t)) - C_I(t,e(t)) - C_{O\&M}(t,e(t)) - C_L(t,e(t)) - H(t,e(t))) \cdot IC_t \cdot D_t$$

$$(10.5)$$

where

$C_{\text{Inc}}(t,e(t))$—income from sales of energy;
$C_I(t,e(t))$—annual expenses on capital investments;
$C_{O\&M}(t,e(t))$—annual, independent of load, maintenance and operation costs;
$C_L(t,e(t))$—losses annual costs;
$H(t,e(t))$—annual penalty function due to technical limitations non-observance;
IC_t—inflation coefficient;
D_t—discounting coefficient

Technical limitations are observed within the entire optimization process. Specifically, observing admissible voltage losses ΔU_{kr} and admissible power flows P_{lkr} we have

$$H(t, e(t)) = H_1(t, e(t)) + H_2(t, e(t)), \qquad (10.6)$$

where

$$H_1(t, e(t)) = \sum_{i \in M} A \cdot P(i) \cdot (\Delta U(i) - \Delta U_{kr})^{\alpha}, \qquad (10.7)$$

$$H_2(t, e(t)) = \sum_{j \in N} B \cdot (P(j) - P_{lkr}(j))^{\beta}, \qquad (10.8)$$

A, B, α and β are empiric constants, which provide penalty function form, M—set of nodes with non-admissible voltage losses and N—set of links with non-admissible flows.

Limitations are also on capital investments in a step if the maximal capital investments are limited.

4th operation Functional $f(t,e)$ is calculated by formula 10.4.

5th operation The membership of D-state with m realized D-actions $e(t,m)$ in OIS set is verified with functional $\max Ef(t,e(t))$ value

$$\max Ef(t, e(t)) = \max_{e(t,m) \subseteq e(t,m-1)} [f(t, e(t,m)) - f(t, e(t, m - 1))]. \qquad (10.9)$$

D-state $e(t,m)$ belongs to OIS set only if

$$\max Ef(t, e(t)) \geq 0, \qquad (10.10)$$

i.e., if functional value reduces.

In case of approximate optimization

$$\max \; Ef(t, e(t)) \geq \delta, \tag{10.11}$$

where δ is the maximal effect function relative boundary value $\delta > 0$.

10.3.4 Sub-block Quasidynamic Optimization

Quasidynamic optimization is approximate dynamic optimization with limited OIS set (see Sect. 7.2.1). It is assumed that admissible number of initial states in set is equal to 1. As a result of quasidynamic optimization, only one D-plan is obtained. Applying quasidynamic optimization, less time is consumed but determination of optimal D-plan is not provided. In case if optimal D-plan demands considerable initial capital investments, which are only paid back after a long time, optimal D-plan in quasidynamic optimization is not obtained.

Quasidynamic optimization block major function is to fast determine optimization area (alternative D-actions).

10.3.5 Optimization Results Analysis

Optimization results are transferred to output menus *Development plans → Optimization results*. As a result of this selection, realized D-actions and their realization year appear.

Economic comparison of best D-plans. Best D-plans economic comparison is performed with menus *Development plans → Optimization results → Total criteria, Development plans → Optimization results → Total criteria → Investments pay-back schedule* or with menus *Development plans → Optimization results → Annual criteria.*

Technical analysis of best D-plans. Best D-plans technical analysis is performed with menus *Electrical calculations → Flows* and other related calculations (see Sect. 10.1), indicating with optimization formed D-plan, plans D-step and operational state.

10.3.6 Optimization Results and Decision Making Under Information Uncertainty Conditions

Optimization algorithm under information uncertainty conditions is shown in Fig. 10.29. Optimization under information uncertainty conditions occurs with active participation of user (engineer analyst). Before optimization, analyst together with expert group must estimate probable information uncertainty range; possible forecasts must be worked out defining credibility level (weight).

Fig. 10.29 Optimization
algorithm block-diagram
under information uncertainty
conditions

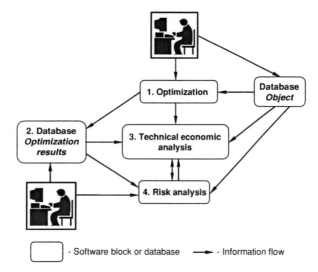

- Software block or database   ⟶ - Information flow

A forecast with higher credibility level shall be entered in the database. Possible shorter planning period must be selected. Optimization is going on with database. Optimization results are as follows: optimal D-plan and competitive D-plans are input into block *Development plans model*. After optimization program, go to the block *Technical economic analysis of assigned development plans*, operating with base data and with D-plans selected in optimization. Such operation scheme is also valid in case of unambiguous information.

In case of uncertain information (there is a set of possible forecasts), it is necessary to proceed to risk analysis with user participation.

Prior to risk analysis, user can correct the D-plan set determined in optimization: exclude non-valid D-plans and add his own D-plans. With such D-plan set, block *Risk analysis* is operating. Forecasts are input by user in the process of analysis. With current forecast and D-plans set from block *Development plans model* program, go to the block *Technical economic analysis of assigned development plans*, from which interim results can be obtained. After all forecasts are reviewed, risk analysis results are output. During risk analysis, D-plans adaptation is performed (approximate optimization, see Sect. 7.4) for the conditions which are determined by forecast when optimizing D-plans in adaptation period with recursive equation 7.14.

In case of necessity, risk analysis can be continued adding supplementary forecasts. Formation of forecasts can also be started anew again.

10.3.7 Sub-block Risk Analysis

In sub-block *Risk analysis*, applied methodology is reviewed in Sect. 1.3. Risk analysis flowchart is shown in Fig. 10.30.

Fig. 10.30 Flowchart of risk analysis; N number of forecasts

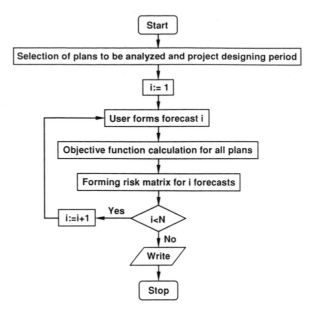

Fig. 10.30 Flowchart of risk analysis; N number of forecasts

Sub-block *Risk analysis* is able to operate with both user's formed D-plans (F), and with optimization software formed D-plans (O), therefore, prior to risk analysis, it must be indicated to what D-plan set (F or O) it will be applied. Depending on user intention, it is possible to change designing period too. This large formation is done prior to cycle entering by forecast. According to our experience, it is not possible to anticipate all forecasts for consideration, thus, information technology provides a possibility of terminating formation of forecasts and renewing it after additional analysis.

Forecast is characterized by:

- Forecast credibility (weight),
- Load increase coefficient by steps,
- Losses price increase coefficient by steps,
- Interest rate in percentage,
- Inflation percentage rate for investments and expenses in network,
- D-actions cost,
- Generation (electric power plants) parameters (power and electricity sales price).

The risk analysis results are filed in table format (see Table 4.3). It is advisable to perform analysis only for technically valid D-plans.

Selecting menu *Analysis* from the main menu, it is possible to access sub-block *Risk analysis* (uncertainty). Selection structure is represented in Fig. 10.31.

Development plans sensitivity analysis and *Risk analysis* calculates technical economic figures for each D-plan, observing various forecasts of development conditions, which allow making decisions under information uncertainty conditions.

Fig. 10.31 Structure of *Uncertainty analysis* block

Fig. 10.32 Forecast formation: start of *Risk analysis*

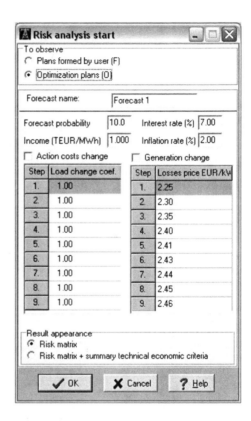

Development plans sensitivity analysis is discussed in Sect. 10.2.3. During the analysis, user can operatively change loads, D-actions and losses costs, as well as interest rate. Performing these changes, it must be taken into consideration that initial input data will remain in the databases. Changes performed in risk analysis will only be related to the particular forecast.

Load and losses costs are changed in proportion to attributed data in the database for each D-step. Due to that, one can:

1. Fast estimate the influence on technical economic results of several parameters simultaneously,
2. Select the best D-plan by minimal risk in case of information uncertainty.

Fig. 10.33 Forecast formation: *Risk analysis* follow-on

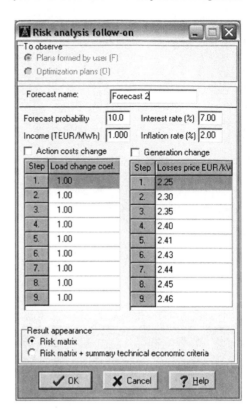

Fig. 10.34 D-actions costs
shifting window

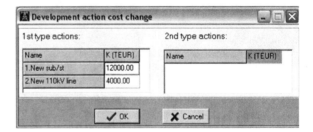

It is recommended that risk analysis be performed after more valid and
equivalent D-plans are determined; and it must be estimated which of them is the
most optimal considering information uncertainty.

Start of risk analysis means that the program is forming first forecast (if forecasts have been already formed, then such are deleted). Risk analysis is continued
when the first forecast is already formed with risk analysis initial stage. In this
way, up to ten forecasts can be formed, which are compared in the summary table.
Either optimization software or D-plans formed by user are analyzed. Risk analysis formation is illustrated in Figs. 10.32, 10.33, 10.34 and 10.35.

Fig. 10.35 Generation
parameters replacement menu

Code	Name	Pnom (MW)	Cdg.(EUR/10kWh(t))
302	BIL	10000.00	2.50
308	RTEC-1	129.00	1.00
310	RTEC-2	815.00	0.40
312	RHES	402.00	0.10
314	Keg1	100.00	0.50
315	Keg2	162.00	0.50
316	PlavHES	847.00	0.10
317		1300.00	4.50
327	IgaunES	200.00	2.00
328	BaltES	180.00	1.90
329	Ignalina3	300.00	0.30
335	LietES	1800.00	2.20
338	KrHAES	225.00	2.30
352	SovetskES	1000.00	2.50

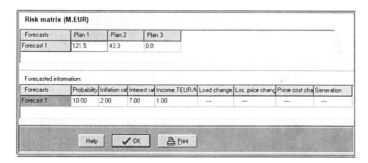

Fig. 10.36 Forecast formation: *Risk analysis* results

Risk analysis results are shown in Fig. 10.36.

Changes can be done for one D-action, for several D-actions or also, in case of necessity, for all D-actions (see Fig. 10.34). Similarly, generation parameters can be changed for one or several electric power plants, or, if necessary, for all electric power plants (see Fig. 10.35).

Figure 10.36 shows risk matrix after calculation of the first forecast. Proceeding with forecast formation, risk matrix is automatically supplemented.

Conclusion

The original methods reviewed in the book are elaborated on the basis of exact methods integration with using optimized systems development regularities. It is relevant to application and utilization of accumulated experience and expertise for reduction of search area. Therefore, the basic principle of artificial intelligence is applied: by investigating experience available, to draw conclusions, on whose basis to make decisions how to manage activities. Certainly, the methods developed according to such principles, are not able to exclude hundred percent of errors. Although a certain risk exists, these methods are still able to solve tasks which cannot be solved by other methods. The laboratory paid much attention also to approximate methods which still give better solutions because much more factors can be observed.

Part I (Chaps. 1–3) evidently demonstrates that development management is not a single event that solves all problems for a long time; instead, it is an uninterrupted multi-step optimization process in which consequences of the decisions made for today or near perspective must be evaluated for long-term perspective. The authors believe that this approach is also valid for many and various systems. Still, it cannot be affirmed that all problems reviewed in Part I have been investigated for all systems. Research has to be continued on management criteria, anticipated perspective duration selection, risk analysis methods, as well as other related aspects.

Part II (Chaps. 4–7) addresses the major methodological and algorithmic as-pects of the development optimisation (OIS) methods elaborated. The methods are based on both analytical and experimental research, and on the long-term practical application experience analysis. In the authors' opinion the material discussed in this part can be viewed as an abstract theory for mathematical systems analysis that is independent of a specific system. Due to that, it can serve as a theoretical basis for elaborating information technology for diverse LTS sustainable development management. The development process modelling and optimisation methods proposed in the book can be successfully applied to meet today's challenging conditions.

Part III (Chaps. 8–10) illustrates practical application of theoretical methods in electric power system management. However, it should be noted that information technology construction principles reviewed in this part can be applied di-rectly only to energy power systems. For other large technical systems, the mate-rial of Part III can serve as a basis for performing a new research aimed at elaborating appropriate information technologies.

The authors deem that the book will achieve its aim if it gives an impulse to new ideas for researchers and engineers and helps in solving both practical and theoretical problems. The issue of LTS sustainable development management can be evidently attributed to the list of problems that are to be solved to avoid hazardous or catastrophic consequences of currently made decisions in the future.

Index

Breinigsville, PA USA
12 January 2011
253150BV00006B/39/P